MATHEMATICAL METHODS FOR INDUSTRIAL PROBLEMS

MATHEMATICAL METHODS FOR INDUSTRIAL PROBLEMS

Proceedings of the International Workshop held in Tecnopolis, Bari, Italy

September 3-5, 1988

Professor V. Capasso
Dr. R. Caselli

SASIAM
Tecnopolis CSATA Novus Ortus
70100 Valenzano (Bari), Italy

///VSP///
1990
Utrecht, The Netherlands

VSP BV
P.O.Box 346
3700 AH Zeist
The Netherlands

ISBN 90-6764-122-7

CIP-DATA KONINKLIJKE BIBLIOTHEEK, DEN HAAG

Mathematical methods for industrial problems/[ed.] V. Capasso, R. Caselli. - Zeist: VSP
Proceedings of the International Workshop held in Tecnopolis, Bari, Italy, September 3-5, 1988.
ISBN 90-6764-122-7 bound
SISO 642.1 UDC 51-74
Subject heading: industrial mathematics.

This publication was composed by the Electronic Publishing Laboratory of Tecnopolis CSATA Novus Ortus, Valenzano (Bari), Italy

Printed in the Netherlands

CONTENTS

Preface

The role of mathematics as a main tool for development of science and technology has always been quite clear. In ancient times the development of mathematics was, in fact, strictly related to the needs of engineering.

In more recent times the explosion of physical knowledge was strictly connected to the development of adequate mathematical tools. Newton and Leibnitz in the 17th century started the first "computing revolution" by developing the infinitesimal calculus, anticipating thus the Industrial Revolution of the 19th century.

At the beginning of this century great attention was paid to the mathematization of all natural phenomena: Biology, Medicine, Chemistry were not extraneous to this (and in the country of Vito Volterra we cannot forget his great contributions).

But for reasons which are quite obscure to us, after World War II schools of mathematics tending to abstraction and formalism prevailed.

Possibly, the need of a more profound analysis of the foundations of the basic mathemaicial structures was related to the actual incapability of Mathematicians to answer the concrete needs of computation of highly nonlinear phenomena, whose complexity could not be held by paper and pencil.

Fortunately in the 70's we have assisted to a new "computing revolution" as a consequence of the "electronic revolution" which has allowed the construction of computers of high power in affordable dimensions and costs.

More than the concept of the electronic computer, already available after World War II, it has been its miniaturization that has allowed the spread of computing power in every home.

Today, computer simulation is part of the first games of any child. Complex systems are not anymore a taboo, however complex the system can be, and a new stimulus has reached mathematicians to face real phenomena.

Engineering, Chemistry, etc. are again a source of very stimulating scientific challenges, since computer simulation needs anyway first to go through the mathematization of the processes and systems to be simulated.

On the other hand, the same design of more and more powerful computers, robots, flexible machine systems etc., pose interesting and important problems to mathematicians, thus making nowadays as ever mathematics and computers as a whole, the most powerful tools for the development of technology.

To explore this role, an International Workshop on the Mathematical Methods for Industrial Problems was held at SASIAM (School for Advanced Studies in Industrial and Applied Mathematics) in Tecnopolis, Bari, Italy during September 3-5 1988, under the following Scientific Committee:

* R. Borrelli
  Harvey Mudd College - Claremont (CA), USA

* V. Capasso
  SASIAM - Tecnopolis - Valenzano (Bari), Italy

* C. Cercignani
  Politecnico di Milano - Italy

* David G. Crighton
  University of Cambridge - UK

* Bruno Forte
  University of Waterloo - Ontario, Canada

* Jacques Louis Lions
  College de France - Paris, France

The scope was to exchange ideas and discuss problems encountered in the formulation, analysis and computer simulation of mathematical models of industrial processes. Problems of validation of models vs experimental data were also considered.

The program consisted of invited talks by lecturers from universities and research centres involved in university-industry collaboration in the mathematical sciences.

Experts from industries gave examples of mathematical problems arising in industry and the methods employed to solve them.

About fifteen delegates represented the major Schools of Industrial Mathematics of the World and about ten speakers from Italian industries spoke about their own experience of Mathematics in Industry. Including the speakers, over 80 participants attended the workshop representing 10 countries.

Great attention was paid by both the academic and industrial world to the workshop. The Italian National Research Council (C.N.R.) was represented by the President of the National Committee for the Mathematical Sciences, Professor Carlo Ciliberto, also President of the University of Naples. The University of Bari was represented by its President, Professor Attilio Alto, and by the Deans of the Faculties of Engineering and of Science.

SIAM was also represented, by its President Elect, Professor Ivar Stakgold, who, as you can read later, discussed the meaning of Industrial Mathematics.

The opening welcome was given by the President of Tecnopolis CSATA Novus Ortus, Professor Ing. Gianfranco Dioguardi.

Altogether we think that as the first workshop of this kind held in Italy, it has provided a clearer understanding of the potentialities of mathematics as a key tool for technological innovations.

V. CAPASSO

List of authors

G. Alberti	LABEN S.p.A. - SS. Padana Superiore 290, 20090 Vimodrone (Mi), Italy
F. Bevilacqua	AERITALIA S.p.A., C.so Marche, 41, 10146 Torino, Italy
M. Biserni	LABEN S.p.A. - SS. Padana Superiore 290, 20090 Vimodrone (Mi), Italy
M. Bistolfi	ENIRICERCHE S.p.A. - 20097 S. Donato Milanese (Mi), Italy
P. Bologna	FIAR S.p.A. - Via 1 Maggio, 41 - 20021 Baranzate di Bollate (Mi), Italy
V. Capasso	Dipartimento di Matematica, Universitá di Bari - Campus 70125 Bari, Italy
A. Cella	ESPRI - MARC S.r.L., P.zza R. Rossetti, 4 int. 4, 16129 Genova, Italy
A. Colombo	LABEN S.p.A. - SS. Padana Superiore 290, 20090 Vimodrone (Mi), Italy
D.G. Crighton	Department of Applied Mathematics & Theoretical Physics, University of Cambridge, Cambridge CB3 9WE, UK
E. Cumberbatch	Department of Mathematics, Claremont Graduate School - Claremont, CA 91711, USA
B. Forte	Department of Applied Mathematics, University of Waterloo, Waterloo, Ontario, Canada N2L 3G1
B.A. Fusaro	Department of Mathematical Sciences, Salisbury State University, Salisbury, MD 21801, USA
F. Kokert	Institut fur Mathematik - Johannes Kepler Universitat Linz, 4040 Austria
J.L. Lions	College de France, 11 Place Marcelin Berthelot 75231, Paris 5, France
C. Maccone	AERITALIA S.p.A., C.so Marche, 41, 10146 Torino, Italy
G. Magnani	FIAR S.p.A. - Via 1 Maggio, 41 - 20021 Barnazate di Bollate (Mi), Italy
S. McKee	Department of Mathematics, University of Strathclyde, Glasgow G1 1XH, UK
F. Nava Mambretti	FIRESTONE - International Technical Center, Casella Postale 2452, 00100 Roma, Italy

H. Neunzert	Laboratory Technomathematik, Universitat Kaiserslautern, 6750 Kaiserslautern, West Germany
G. Noberasco	ESPIR - MARC S.r.L., P.zza R. Rossetti, 4 int. 4, 16129 Genova, Italy
J.R. Ockendon	Mathematical Institute, 24-29 St. Giles, Oxford OX1 3LB, UK
M. Orsucci	TELETTRA S.p.A., Via Trento 30, 20059 Vimercate (Mi), Italy
L. Peer	Institut fur Mathematik, Johannes Kepler Universitat Linz, 4040 Austria
M. Primicerio	Facoltá di Scienze Matematiche, Fisiche e Naturali, Universitá degli Studi di Firenze, 50100 Firenze, Italy
V. Recchi	TECNARS S.p.A., Via Pionieri del Commercio Barese, Zona Industriale, 70123 Bari, Italy
M. Robin	INRIA, Domaine de Voluceau - Rocquencourt, B.P. 105, 78153 Le Chesnay Cedex, France
G. Soave	ENIRICERCHE, S.p.A. 20097 S. Donato Milanese (Mi), Italy
E. Spedicato	Dipartimento di Matematica, Statistica, Informatica e Applicazioni, Universitá di Bergamo, 24100 Bergamo, Italy
I. Stakgold	SIAM, Department of Mathematical Sciences, University of Delaware, Newark, DE 19716, USA
P.P. Strona	CENTRO RICERCHE FIAT, Strada Torino 50, 10043 Orbassano (To), Italy
H. Wacker	Institut fur Mathematik, Johannes Kepler Universitat Linz, 4040 Austria
J.M. Zijm	Nederlandse Philips Bedrijven BV, Centre for Quantitative Methods, 5600 MD Eindhoven, The Netherlands

Part One: Invited Speakers

1. The SASIAM Experience of Mathematics in Industry

V. Capasso

SASIAM - Tecnopolis CSATA Novus Ortus

Valenzano (Bari), Italy

1.1 Introduction

The School for Advanced Studies in Industrial and Applied Mathematics (SASIAM) has been designed during 1984 in coincidence with the opening of the technopolis called TECNOPOLIS Novus Ortus created in the neighbourhood of Bari, (Italy).

The explosion of the "electronic revolution" in the last 20 years has induced the need of engineers, or more generally scientific, graduates. But the differential development of the northern and southern part of Italy has accentuated the emigration of qualified people from the south to the north.

The main principle of the new plan for the development of Southern Italy is based on the idea that high tech industries should be created, supported by a stronger interaction between the local universities and entrepreneurs.

The specific role of the technopolis aims to establish cooperative training, and research and development programmes with national companies to increase and diffuse technological services for small and medium sized firms and stimulate new entrepreneurial initiatives.
More specifically the programme in the next three years intends to stimulate the creation or anyway the expansion of high tech companies; the implementation of interdisciplinary graduate or postgraduate educational programmes; initiatives which tend to include Southern Italy in the framework of the EEC education, technology

transfer and strategic research activities; initiatives oriented to establish forms of cooperation with developing countries.

During 1984 a large discussion had arisen about the role of Mathematics in the development of the high technology. In the United States the famous "David Report" was published by the Ad Hoc Committee for Renewing U.S. Mathematics of the U.S. National Research Council. Among the others the following statement has to be considered of an historical value "The emergence of 'high technology' brought our society into an era of mathematical technology in which mathematics and engineering interact in new ways....Mathematics and engineering now interact directly on a broader, deeper scale greatly to the benefit of both fields, and to technology" ["David Report" AMS Notices 31(5) (1984)]. One year later other two important contributions came to reinforce the validity of the equation

$$\text{High Tech.} = \text{Math Tech.}$$

From R. Mc Kelvey, Chairman of the Committee for the Status of the Profession [SIAM News Sept.1985] we hear that "In the next few years education in mathematics will face the challenge of a period of fast transformation of knowledge and technology....It must take into account the dissemination of the mathematical way of thinking in wider and wider areas, since the 'mathematization' and subsequent 'experimentation' with computers will penetrate more deeply into our scientific, technological, and commercial activities".

The President of the Mathematical Association of America, L.A. Steen [AMS Notices, 32(4) 1985] states furthermore that "mathematics itself is changing dramatically - in content, scope and application - the truth is that mathematics is not just being applied, but is being continually created in response to challenges from science, from technology.....".

Actually the need of a stronger interaction between Mathematics and Industry had already been felt before by various centers, and here we shall not forget Claremont in California, Oxford in England, among others, which started in the late 60's programmes of interaction university-industry; their experience is described in this volume by the main protagonists.

Anyway this increasing demand on Mathematics has stimulated the founding, worldwide, of a number of academic groups committed to interaction with industry.

Within Europe, after the first European Symposium on Mathematics in Industry (ESMI, Amsterdam 1984), a number of these groups have accepted their responsibility toward European Industry by founding a European Consortium for Mathematics in Industry (ECMI).

The SASIAM experience has the specific peculiarity to have been started in the absence of any previous coordinated interaction of the local universities with industry, even if various interests in applied mathematics already existed. The main difficulty in the interaction between university and the local, even large, industries is due to the fact that usually the companies established in the south have the "brain" elsewhere.

As stated before the TECNOPOLIS Novus Ortus development Programme has the ambition to invert this tendency and favor the creation of either new high tech companies or branches of the R&D centers of companies already existing anywhere.

1.2 Areas of Interest of SASIAM

The specific role of the School, as we shall stress later, is that of establishing a link between the centers of production of the "know-how" in mathematics, statistics and computer science (typically universities and the National Research Council) and the industrial world.

As a consequence, the areas of interest emphasize this role. In the fundamental areas, "general purpose" disciplines are identified, which provide the tools for the simulation of industrial processes, which show particular characters of complexity. Such competences usually are not of a direct interest for the academic centres, being themselves more involved in the development of specific segments of knowledge, more than in the operational implementation of the knowledge itself.

On the other hand, as far as the applied areas are concerned, the School tends to devote most of its interest to real industrial projects, with a particular attention to areas which are complementary or otherwise a support to the other activities already present in TECNOPOLIS.

The School should aim in fact to become a centre of excellence in specific areas of technological interest.

In addition to a set of fundamental areas such as Mathematical Modelling, Computer Simulation, Optimization and Control, Data Analysis, etc... a special attention will be devoted to scientific computing in the following areas:

- Complex Systems
- Large Scale Computing
- Space Sector
- Telecommunication
- Reliability and Quality Control
- Image Processing
- Geometric Methods
- CAD/CAM
- Operation Research and Optimization
- etc.

1.3 Programmes of International Cooperation

The School operates in strict contact with similar institutions of universities and research centres in Europe and North America (USA and Canada) in order to create more cultural and scientific cooperations.

In particular, in the context of the ECMI, the School participates together with the Universities of Eindhoven (NL), Kaiserslautern (FRG), Oxford (GB) and Linz (Austria) in a joint postgraduate educational programme.

The area of applicable mathematics is so large and diverse in knowledge and techniques that no single center or even country can possibly claim to be able to provide all the expertise needed. We are working hard so that other centers in Europe will participate to such a programme and in fact we also cooperate in the framework of ECMI in many different ways with:

- Rolf Nevanlinna Institute
  Helsinki - Finland
- Centre de Mathematiques Appliquées
  Ecole Normale Superieure
  Paris, France
- INRIA
  Rocquencourt - France
- Dept. of Mathematics
  University of Strathclyde
  Glasgow, U.K.
- Dept. of Mathematics
  National Institute for
  Higher Education
  Limerick, Ireland
- Istituto Matematico "U. Dini"
  Universita' di Firenze
  Firenze Italia
- CWI
  Amsterdam - The Netherlands
- Division of Mathematical Sciences
  Norwegian Institute of Technology
  Trondheim - NTH - Norway

Within ECMI a joint Certificate of Expertise in Industrial Mathematics has been designed, supported by the EEC programmes COMETT and ERASMUS.

More in general the international cooperation is based on professor and student exchange programmes; the joint supervision of theses and work-groups on industrial projects; reciprocal acknowledgement of credits.

1.4 Cooperation with Industry

1.4.1 The "MATH CLINIC"

In order to allow students to experience directly the application of mathematics by formulating, analysing and possibly solving problems submitted by industries and public institutions, SASIAM has introduced an activity of "Math Clinic" according to a scheme already acting successfully for more than ten years in those universities who cooperated for the establishment and development of the School, such as Claremont (Ca,USA), Waterloo (Canada),etc.

As you will read elsewhere in this volume the "Clinic" activity requires the students to be directly involved in the solution of typical industrial problems through the direct participation in a project submitted by an industry.

Any project is developed by no more than three or four students under the supervision of an Advisory Committee so organized:

- an instructor of the School;
- an expert from the industry submitting the problem;
- a tutor with a successful experience in a similar project in one of the universities cooperating with SASIAM.

One of the main roles is played by the industry liaison who takes part in the work group. The information exchange which can be realized this way is very important especially at the beginning because it helps the team to focus on the primary objective of the project.

It is understood that confidentiality is assured about data and results for at least one year from the conclusion of the project (with the final technical report).

1.4.2 Benefits for the Industrial Partners

1. A view on new important research
2. Participation in R & D projects
3. Access to high quality students
4. Financial benefits
5. Competition benefits
6. Progress of technological research

1.4.3 Club of Companies

The School intends to establish a network of companies interested in the applications of mathematics with a data bank of all the problems submitted by industries and their possible solutions.

1.4.4 Industrial Monitors

The link with the industry will be supported, on an operating basis, by "Industrial Monitors": researchers and consultants of companies participating in the SASIAM programme who will maintain an informal but constant connection with respect to the research projects submitted to the work teams of the School.

1.5 Educational Activity

1.5.1 Postgraduate Programme in Industrial Mathematics

Tuned within ECMI,the School offers a programme to train graduates in scientific areas to become professional applied mathematicians, and is intended for those who wish to prepare themselves for employment in industry, government and other applied research centres.

Students in the programme are expected to acquire a broad knowledge of the mathematical sciences, together with experience in relating to the mathematical needs of the nonacademic work environment.

The School is organized on the basis of a series of courses, plus the participation to industrial projects, within two-three years.

1.5.2 Special Courses

The School also offers a programme of "Continuing Education" based on a series of special courses of highly specialized nature.

Each special course is delivered by experts of international reputation in a week with an intense laboratory activity and "case studies".

The courses are designed for researchers and technicians already involved in research activities and/or working in industry.

1.5.3 Colloquium

In order to improve the exchange of scientific knowledge between academic institutions and the industrial environment on the role of mathematics in industrial production, the School offers a series of colloquium talks delivered by international experts who are responsibles of research centres in industrial and applied mathematics.

1.6 Editorial Activity

Great attention is payed to the diffusion of know-how in the area of applied and industrial mathematics. To this end the School stimulates all invited speakers and lecturers to produce a set of lecture notes. To speed up the circulation they appear at first as SASIAM Reports even in a draft format, which may be later published in a more accurate form in a SASIAM series of Lecture Notes.

1.7 Laboratory

The complexity of the tasks which are the objective of both the research and the educational activities of the School needs advanced computing tools capable to manage specific software products which are often available at an experimental stage.

Particular attention is paid to the acquisition of advanced hardware as well as software instruments which allow us to operate in the fields of simulation of complex systems, of symbolic computation, of high resolution 3D graphics.

In addition, considering the need of high quality post- processing applications oriented to the animation of the displayed images coming from simulation, the School is going to be equipped with suitable HW/SW tools, which will include acquiring and displaying high resolution images.

A suitable series of workstations with or without local intelligence permits the access to the computing resources already available in TECNOPOLIS as well as to advanced structures in other international centers.

1.8 Software Library

An important objective of SASIAM is the implementation of a library of software dedicated to industrial applications.

1.9 Workshops

On a yearly basis SASIAM shall organize a workshop on the "Role of Mathematics in Industry" like the present one. The School may also host initiatives of the same kind by other groups or organizations.

1.10 Project Cases at SASIAM

To illustrate some of the projects on which SASIAM has collaborated with industry, three brief descriptions are presented here.

1.10.1 Image Processing

A digitized image of a real world black and white picture is a finite set of "pixels" (pictorial element). These pixels form an $m \times n$ matrix and to each of them a natural number g , representing the gray level, is attached. Usually the number g varies in the range (0, 255) but in many applications, specially in computer vision and robotics, for which transmission and storage of images are required, may be necessary to deal with a reduced number of gray levels. This can be obtained by thresholding. Clearly this yields a loss of information on the subject. For example in the binarization process (reduction to two grey levels) if we denote by g any grey level between black (g_0) and white (g_l), the thresholding will transform g as follows.

$$g \rightarrow g_0, \ \text{if } g \leq t^*$$

$$g \rightarrow g_l, \ \text{if } g > t^*$$

The loss of information depends on the choice of the threshold values t^* . It seems to be natural to choose such value so as to minimize the information loss. A new method for optimal thresholding by minimum information loss has been recently [2] introduced. The method provides interesting improvements on previous experimental results.

In [1] it is first derived from a "natural" set of axiomatic properties the general mathematical representation of a measure of the information loss. Then it is provided reasons for the recourse to a formula to measure the loss of information which is more sophisticated than the one used in [2].

In a global point dependent technique to threshold a digitized image the only "relevant information" about the subject is its grey levels histogram. In fact one replaces the image function $g(i, j)$ with the probability distribution of the grey level

$$p(g_t) := probability \{g = g_t\}, \quad t = 0, 1, \ldots, l.$$

Any reasonable measure of the amount of uncertainty U, that one has about the subject of an image, given its grey level histogram, should be a real number that takes into account this "relevant information". That is U should be a real valued function of the grey levels

$$\{g_t, \quad t = 0, 1, 2, \ldots, l\}$$

and their probabilities

$$\{p_t := p(g_t), \quad t = 0, 1, 2, \ldots, l\}$$

On the basis of the set of axiomatic properties introduced in [1], the following (unique) representation for the measure of uncertainty (entropy) has been derived:

$$\begin{aligned} U = & -Ap_0 \ln(p_0) \\ & -A\sum_{i=1}^{l} p_i \ln(p_i/(g_i - g_{i-1})) \\ & -\sum_{i=1}^{l} f(g_i - g_{i-1}) + \varphi^* \end{aligned}$$

where A is a nonnegative real constant, φ^* a real constant and f a function $f: R \to R$.

Let's indicate with:

$U =$ the measure of uncertainty about the subject given the histogram of the digitized image

and with

$U' =$ the measure of uncertainty about the subject given the histogram of the binarized image.

The threshold value t^* for the binarization process is the value which minimized the loss of information $U' - U$. The results obtained show that the choice

$$f(u) = h_0 u$$

with h_0 a real positive constant, may lead to unsatisfactory binarizations, while the choice

$$f(u) = h_0 u + h_1 u^2$$

with both h_0 and h_1 positive constants, at least eliminates the inconsistencies found before and provides significant improvements in a large variety of real world images, mainly for picture with bimodal grey level histogram.

Experimental results has been obtained also with the so called "modified method" that is based on the remark that in binarizing an image those pixels that are already black should stay black and the white ones should stay white.

1.10.2 Spatial Data Analysis

A problem of great importance in the statistical analysis of bidimensional spatial data concerns the stochastic independence of the two positional coordinates of the data themselves. The objective of this study is just to establish the degree of stochastic independence of the two random variables coordinates on the basis of their pictorial representation. Theoretically the method that it has been used, called Maximum Expected Information (MEI) Discretization, is based on the principle of maximum entropy as a criterion for optimality of expected information.
Two possible applications of the above mentioned method are one related to town - planning, where the spatial pattern of retail shops in urban areas has been analysed, and one in the biomedical area where the distribution of a disease with respect to the age of the patients has been studied. Another application of great importance, is related to the simulation of space - time stochastic systems, such as the statistical analysis of pseudo - random number generators.

1.10.3 Thermal Treatment of Special Steels

In order to improve the mechanical properties of steel, an already shaped block of this material must be submitted to a thermal treatment based on the gradual raising of the temperature up to a certain predetermined level, and a subsequent permanence of the block at that level.

The mathematical problem is that of optimizing the time elapsed in the raising of the temperature from the environment level to the final level, in such a way that the steel is not subject to physico-chemical stresses due to high gradients of temperature in the block itself. On the other hand a prolonged time for the raise of temperature, in addition to energy consumption, may induce undesirable chemical modifications of the material,which may affect its mechanical qualities.

The mathematical interest relies in the fact that the system is based on nonlinear parabolic equations in domains of complex geometry. Since some of the parameters are unknown, an identification problem also arises, on the basis of experimental data.

Due to the nonlinearity of the equations and the complex geometry of the system, the computer simulation, based on 3D finite elements, also poses stimulating problems.

1.11 Acknowledgements

It is a great pleasure to acknowledge FORMEZ (Centro di Formazione e Studi per il Mezzogiorno), Regione Puglia, the National Research Council of Italy (CNR)-National Committee for the Mathematical Sciences, the E.C. plans COMETT and ERASMUS, and Tecnopolis Csata Novus Ortus for their support to the School.
The projects presented in 1.10, "Project Cases at SASIAM" have been carried out in collaboration with IBM- Italia and Breda Fucine Meridionali S.p.A.

1.12 References

[1] R. Caselli and B. Forte, "Thresholding Grey Level Histogram by Minimum Information Loss", preprint.

[2] B. Forte and P.K. Sahoo, "Minimal Loss of Information and Optimal Thresholds for Digital Images" - Computer Vision, Graphics and Image Processing, to appear.

2. Mathematical Modelling and Asymptotic Analysis in Engineering Acoustics

Prof. D. G. Crighton

Department of Applied Mathematics & Theoretical Physics

University of Cambridge, U.K.

Modern acoustics provides many problems in which mathematical modelling and asymptotic analysis are very important, both for understanding (and control) of the physical processes and for numerical predictions to be used in preliminary engineering design (followed by full computational solutions *only after* the important features have been extracted using asymptotic techniques).

Here we give three examples of this philosophy. In each case one has to understand that the acoustic fields generated in air or water, even by very powerful machines, have absolutely negligible energy compared with the main process (e.g. propulsion of an aircraft or ship) for which the machine is designed. For example, the sound made by an early (1960) jet transport aircraft is louder than the sound of the population of the whole world shouting together; but the energy in that sound, continuing for the one minute of take-off, is only just enough to cook one egg! With a modern jet transport (e.g. Airbus A310) the sound is much less, only equal to the sound of the people in a large city shouting together, and the sound would have to continue for one day to get enough energy to cook one egg! In the case of ships and submarines, sound is generated in the water, but this is much smaller still (than in aircraft) in relation to the propulsive energy of the motors, or the energy needed to overcome the surface wave resistance of a ship. The sound can easily be detected by sonar receivers, but the sound energy is smaller than the energy in a ship wake (for example) by perhaps a factor of 10^{-15}.

Clearly, no direct computational codes can accurately calculate the sound. Even numerical round-off error in the computation of the flow of air round an aircraft or water round a ship can act as an acoustic source which is much more powerful than the "real" acoustic sources (the highly turbulent exhaust stream from a jet engine, or the highly turbulent wake of a ship, say).

The aim of the mathematical modeller must be to extract exactly those parts of a very complicated turbulent motion which generate sound, ignoring (or suppressing) those much larger parts which do not. A combination of skill in mathematical modelling and in the use of asymptotic techniques has played a large role, in the last 40 years, in the understanding, prediction and control of jet aircraft noise, underwater noise, and, most recently, the noise of new high-speed propellers ("propfans") to power transport aircraft in the 1990's and next century.

Example 1. Jet noise.

To understand the sound (noise) generated by the turbulent exhaust stream of a jet engine, Lighthill (in 1952) transformed the very complicated nonlinear equations of fluid dynamics (the Navier-Stokes equations) into the form of a linear sound-wave equation, with a "forcing" or "source" term,

(1) $$Lp = S .$$

L is a simple linear differential operator describing the propagation of sound waves, p is the pressure in the air, and S is a source term confined to the turbulent exhaust region. If S can be predicted or measured, the sound p can be calculated by classical acoustics methods applied to (1). Lighthill manipulated S to extract the factors which make sound generation by turbulence so inefficient. This permits one to at least correctly estimate the order of magnitude of p , but to accurately predict p by numerical calculation or measurement of S is still quite impossible (except in one or two very well ordered simple flows, which can be used as test cases).

The main achievement of the modelling and analysis of (1) is, however, a famous scaling law; the intensity I of sound (proportional to p^2) is proportional to the eighth power of jet exhaust speed U,

(2) $$I \sim U^8 D^2 ,$$

where D is the diameter of the jet nozzle. Since the mechanical power of the engine is proportional to U^3D^2, this means that a large reduction in I can be achieved by a reduction of U and an increase of D to keep the power constant. For example, if U is reduced by a factor 2 and D is increased by a factor 2.8, the sound intensity is reduced by a factor 32. Exploitation of this effect has been carried out on jet engines since about 1960, and has contributed much of the 30 decibel (factor 1000) reduction in jet noise that has been achieved since that time.

The mathematical law (2) has also been very widely used in experiments. Often in experiments on jet engines one finds a variation like U^4 or U^6. It is understood, from the mathematical theory of (1), that such sound cannot be generated by the turbulent exhaust jet, but must come from noise sources within the engine itself (from combustion, rotating machinery, etc.). The law (2) is used as a *definition* of pure turbulent exhaust noise, and used in experiments to identify sources of noise.

There is a very extensive body of work on the Lighthill theory, which has provided most of the essential ideas for understanding and controlling jet noise. It is not a theory which is ever likely to give accurate numerical predictions, but it has provided a rather simple, general, versatile way of thinking about the problem, of interpreting measured data, of suggesting critical experiments, and of achieving large reductions in noise.

Example 2. Underwater noise in bubbly liquids.

In many underwater applications, there are significant concentrations of small air bubbles - as in the case of a ship wake. Even small volume concentrations of bubbles change the compressibility of water (and also the speed of sound) very much. In pure water the sound speed is $c_0 = 1500$ m/s, and in pure air it is $c_a = 340$ m/s, but it does not decrease continuously in a bubbly liquid as the concentration of air is increased from 0% to 100%.

Rather, the sound speed first decreases to small values for quite small concentrations, then increases up to 340 m/s as the concentration approaches 100%. Sound speed c in the mixture can be as low as 100 m/s for a volume concentration of about 1%, and speeds as low as 20 m/s have actually been measured at higher concentrations. This effect is important when we examine the generation of sound in water - for example, by the bubbly turbulent wake of a ship.

The way to do this is to study the Lighthill equation (1) and to extract from the source S a term which is zero when the bubbles are absent but which is large when the sound speed c in the turbulent bubbly region is very different from the sound speed c_0 in the pure water where the observer is listening. It turns out that the result for the intensity of sound can be written

$$I = NI_0\left(\frac{c_0}{c}\right)^4, \tag{3}$$

where I_0 is the intensity generated by the same turbulent flow with no bubbles present, and N is a number which cannot be predicted by the theory, but is expected to be around unity (almost certainly between 0.1 and 10). The big effect is represented by the factor $(c_0/c)^4$, which can represent an increase by a factor of between 10^4 and 10^6 for bubble concentrations of around 1%.

Prediction of this spectacular large effect is another good example of the power of the mathematical modelling of the Lighthill theory. It is extremely unlikely that any computational approach could convincingly and correctly predict this. Similar examples exist, of careful modelling in the Lighthill framework, to predict the sound from dust particles in air, from unsteady burning of a flame, from shock waves passing through turbulence, from the breaking of ocean surface waves, and from many other physical processes.

Example 3. Propfan noise prediction.

Proposals have been made in the last 10 years for high speed advanced propellers to replace jet engines on future transport aircraft. Several prototype propeller engines have now been tested, and it is claimed that they will reduce fuel consumption by between 20% and 50%, compared with jet engines, for the same size and speed of aircraft. A critical legal requirement is that these new aircraft should be very quiet, at least as quiet as those which they replace. The problem of predicting and controlling the noise of high speed propellers turns out to provide good opportunities for mathematical modelling and for asymptotic analysis to simplify the mathematical expressions. The simplified asymptotic expressions give insight into the physical processes, and suggest ways of controlling the noise (without reducing the engine power); and they also give quite accurate numerical predictions, when compared with full-scale experiments and with the results of large-scale computation (which is too expensive and time-consuming to carry out at the preliminary design stage).

Consider a single propeller, with B identical blades set uniformly on a shaft which rotates at a uniform angular velocity Ω. The propeller blades generate moving forces on the air, and also have a moving displacement ("thickness") effect. Both of these cause sound waves to be radiated away, and the sound is periodic with frequencies that are multiples of the blade passage frequency $B\Omega$,

(4) $$p(t) = \sum_{m=-\infty}^{+\infty} P_m \exp(imB\Omega t)\,,$$

where $p(t)$ is the sound wave pressure measured by an observer in a particular direction at a particular distance. We have to predict the harmonic amplitudes for sound travelling in all directions and distances, for a number of different operating conditions of the engines and designs of the blades, and for all frequencies $(mB\Omega)/2\pi$ to which the human ear is sensitive (from 20 cycles/sec. up to 10000 cycles/sec.). This can be done from the laws of classical acoustics if the blade shapes and the blade forces are given, but for each condition and each m, P_m has to be calculated from a complicated double integral over each blade surface. The computational time is enormous, and the results do not suggest any way to control the sound at some particularly annoying frequency.

Modelling and asymptotics provide a much better solution. In the simplest modelling we say that the blades are simply radial lines, rotating and applying force and displacement effects to the air. Then P_m is given essentially by one integral along one blade, from $z = z_0$ say, (the hub), to $z = 1$, (the tip), of the form

(5) $$P_m = \int_{z_0}^{1} S(z) J_{mB}\left(\frac{mBz}{z^*}\right) dz\,,$$

where J is a Bessel function and z^* is called the "Mach radius" (it is a known function of the direction of the observer and of the rotational and axial speeds of the blades). The function $S(z)$ is known, related to the distribution of force and thickness along the blade.

We propose to evaluate (5) approximately by asymptotic techniques based on the idea that the number of blades B is large. In fact what is needed is that mB is large, and with a modern propeller of at least 8 blades, mB is clearly large except when $m=0$ (and when $m=0$ there is a factor m in S which makes this case trivial). Suppose that conditions are such that z^* is greater than 1. Then the Bessel function has its largest value at the tip $z=1$, and decreases exponentially rapidly as z decreases from 1. This allows the integral in (5) to be completely evaluated under the large mB approximation. We get a simple formula for *all* the P_m which can be compared with the results of full numerical computation and is quite accurate enough for all normal engineering purposes (even for mB as small as 4). Further, the result shows that almost all of the sound is generated by the distribution of force and thickness at the blade tips. To reduce the sound one must therefore redesign the blades to reduce the loading and thickness at the tips, and one may hope to do this without making the propeller less powerful or efficient. The largest forces on the blades do not act near the tips, but those large forces ($S(z)$ large in (5)) make very little sound because they are multiplied by a very small Bessel function.

In other conditions z^* may be less than 1. Then the Bessel function reaches its maximum value close to $z=z^*$, and this is the position on the blade which now generates most of the sound. Sources with z less than z^* generate little sound because the Bessel function is small there, sources with z greater than z^* generate little sound because the Bessel function oscillates rapidly there, so that the contributions from $z>z^*$ almost cancel. In fact, when $z^*<1$ the integral (5) can be approximately evaluated by these arguments to give simply

$$P_m = \frac{z^* S(z^*)}{mB} , \tag{6}$$

which is easy to compute quickly for any conditions, and which shows that the dominant sound source is now located at the Mach radius, $z=z^*$, not at the tip, $z=1$. The prediction of (6) agrees quite well with the results of a full numerical calculation, and the agreement is made very close indeed if a higher approximation is calculated. Equation (6) is the first term of a series of which the next term is proportional to $(mB)^{-3/2}$, and represents sound generation by the blade tip. Again we see that the asymptotic predictions are accurate, and reflect the physical processes in a way which should lead to control. For example, the legal certification limits might be exceeded only in some frequency ranges and in some particular directions by some proposed aircraft-engine design. Our theory explains what parts of the blade are responsible for sound at a particular frequency and in a particular direction, and the design engineer must then modify the blades at the appropriate sections.

Acoustics offers many examples like this, where mathematical modelling of complicated processes, followed by the use of asymptotic analysis, gives understanding, accurate predictive capability, and hints for control and improvement. Other examples, including problems in which structural vibration is important as well as noise, are discussed in

2.1 References

[1] D. G. Crighton 1988 "Aeronautical acoustics; mathematics applied to a major industrial problem", Proceedings of ICIAM '87 (First International Conference on Industrial and Applied Mathematics, Paris 1987), SIAM Philadelphia

[2] D. G. Crighton 1989 "Fluid loading : the interaction between sound and vibration". Rayleigh Medal Lecture, Journal of Sound and Vibration, to appear.

3. Case Studies from The Claremont Mathematics Clinic

Ellis Cumberbatch

The Claremont Graduate School

Claremont, CA 91711, U.S.A.

Mathematics students at the Claremont Colleges may elect to take a clinic, as a course for credit, usually in their senior year or in their first or second year of graduate school. Each clinic project is an unsolved mathematically-oriented problem that has arisen in local industry. The industrial sponsor pays a fixed fee (currently $28,000) for the services that the clinic team provides over the academic year. A team consists of a faculty supervisor and four or five students, one of whom is provided with a fellowship to compensate for the responsibility of being team leader. The sponsor also designates a liaison - working scientist or engineer - to monitor the goals of the project and to provide technical knowledge of the processes or products under investigation. The fact that the problem comes from the real world, that the client is willing to pay a substantial fee for its solution, that professional oral and written reports are to be delivered throughout the year, shows to the mathematics students that **mathematics is useful**, and they respond by working hard on these projects, by taking more mathematics courses, and eventually choosing to take up mathematics as a career.

To date there have been more than 80 year projects in the 15 years of our operation - currently we are at steady state with 6 or 7 clinics a year. (For more information see [1] , [2] , [3].) Sponsors include large corporations in the oil, aerospace and electronics industries, but also our local water company, the State Architect, and the U. S. Forest Service. We have had over 30 different clients, so many have sponsored repeat projects, some immediate continuations, some quite changed. The scope has been wide: statistics, scheduling, economics, game theory, computer graphics, numerical analysis, ordinary and partial differential equations, control theory, fluid and solid mechanics, field-effect transistors.

Before looking at four of our recent projects, I wish to make some remarks about the clinic concept, which has been successfully copied elsewhere, and the study group concept started at Oxford 21 years ago, which also has been copied successfully. In addition to providing a valuable consulting service for industry, both these initiatives provide a showcase to advertise that mathematics is an increasingly important tool in the real world, and this information is useful in persuading what has been a declining number of students interested in mathematics to stay in the subject longer and perhaps as a career. The clinic does this directly, as undergraduates get the chance to participate on a real project, not one that is a professorial "thought experiment." Also the professional training, especially the team leader's, provides the experience that employers want in the mathematicians they hire. The study group experience for undergraduates is second-hand, relying on the new insights about modern technological needs provided to faculty, but for advanced graduate students and post-doctoral fellows is immediate and direct, furnishing new phenomena to model and opening up new areas for mathematical analysis (e.g. John Ockendon and Alan Tayler have emphasized in many talks how free boundary problems are so surprisingly dominant in the problems that come to the Oxford Study Group). In sum, the clinic and study group concepts have different goals: the clinic has undergraduate and master's level education as a primary focus, whilst the study group provides new problems for Ph.D. theses and faculty research. I feel fortunate in having had experience of both these initiatives, and I wish good fortune to this new venture at SASIAM which has both clinic, study group and **new** features.

Now to some of our recent clinic projects:

Outgassing and Contamination in Vacuum Systems

Client: Aerojet Electrosystems

Outgassing is a process whereby a contaminant diffuses through a source, desorbs through a boundary, is transported to another site, called the target, and is adsorbed there. The source is a metal, plastic, or coat of paint, containing a concentration of contaminants or generating them due to, say, chemical reactions with ultraviolet radiation. The target may be a satellite optical system, or an electronic component under manufacture. Although there is a considerable amount of published material in this area (e.g. [4] - [9]), the client was still meeting difficulties in identifying sensitive parameters in the problem, and in relating observed contamination of components in the field to experimental results and to various ad hoc theories that attempted to describe this phenomenon.

The approach adopted by the clinic team was to model the physical processes of out-gassing contamination by setting up equations of reaction-diffusion in the source, convection-diffusion or molecular flow in the vacuum, together with rather complicated boundary conditions on the source and target surfaces. Even in the simple plane geometries adopted, the equation set filled a page, and there were 16 physical, geometrical, and initial-value parameters to describe. There is little point in displaying those equations here and the interested reader is referred to [10] , [11]. The team developed various analytical and numerical schemes. The first concerned the "well-mixed" approximation where the diffusion in the bulk is neglected so that time is the only independent variable and the transport processes across the source and target boundaries become the governing control parameters. The second looked at the source boundary taking other boundaries far away, so that exact solutions of the diffusion equation could be written down for the two semi-infinite regions adjacent to this boundary. The third solved the diffusion problems in adjacent finite slabs by a Green's function technique; this allows the fluxes approaching and leaving the source boundary and that approaching the target boundary to be related through a system of three Volterra integral equations. This system was solved numerically and results shown graphically. Subsequently, these equations were shown to represent mathematically well-posed problems, as existence, uniqueness, positivity and some asymptotic behavior results were proved for their solutions, [11].

Optimization of the Use of Mass Spectral Data

Client: Perkin-Elmer, Inc.

The commercial application of the mass spectrometer under study is to monitor for chemical spills in the environment, or in a "clean" factory. A chemical compound is broken into ion fragments as it passes through a spectrometer and the fragments are separated according to their mass-to-charge ratio. The time spent measuring a particular ion count is referred to as the dwell time, and increasing this reduces the statistical noise. There were 43 compounds under investigation, with about 15 of these counted as target compounds, and the rest as potential strangers. The 43 compounds had 113 m/e ratios.

Given a set of target compounds, the goal was to choose the m/e ratios to be measured and to determine the associated dwell times, with a total cycling time kept reasonable, so as to minimize the error of estimates of target concentrations, bearing in mind the sensitivity to compounds not on the list. The reliable detection of strangers, that is compounds not part of the target set, was also an important part of the algorithm.

The general linear statistical model indicates that the smallest variance estimate is given by the weighted least squares technique, where the weights also have to be estimated. Various choices of weights were tried, including constant weights, and weights determined by the alarm level of each compound (where accuracy is most important). A number of simulations were performed and the results compared, showing that weighting is not a significant factor.

The detection of a stranger compound is complicated by the choices of the 113 m/e ratios made for the target set, and the likelihood that the stranger has a spectra which is a linear combination of the spectra of the target compounds. Explanation of these factors is beyond the scope of a review article and we refer to the source publications [12], [13], [14].

Most of the work performed by the clinic teams (two years) was supported by computations done on college machines (IBM and VAX). The client's needs were to translate the programs for use on PC's. Storage requirements inhibited this transfer until the introduction of some faster software (Turbo Pascal).

Development of Mathematical Models of Floating Ring Bearings for High-Speed Turbo Machinery Applications to Analyze Bearing Stability

Client: Garrett Automotive Products Co.

Conventional bearing design has a circular shaft inside a slightly larger circular housing with the shaft supported by the large pressure forces generated in the oil film separating the two components. Various design considerations indicate the usefulness of interposing a sleeve between these two - a floating ring. The goal of this clinic was to organize a suitable mechanical/fluid dynamical model of a floating ring bearing and to examine the resulting equations both for stable and unstable motions. This topic is not new (see [15], [16]), but the geometry is so complicated that the team concluded that it was imperative for understanding to formulate their own model rather than adopt one or other of those already available.

The fluid mechanics of lubrification theory is fairly straight forward, but the resultant forces of two fluid layers acting on solid components with 3 geometric centers which are in relative motion implies the introduction of a number of reference frames. The equations of motion are clearly complicated, and sorting through the possible solutions is an arduous task. Similar to previous work, the team developed two main approximations valid for short and long bearings (short/long in the direction of the generators of the shaft), and carried out numerical simulations showing a variety of motions (see [17]). Various whirling motions could be generated which usually subsided to a limit cycle orbit. Some "chaotic" motions were seen.

The team remarked that a number of authors investigate stability analytically by looking at a perturbation around an equilibrium solution. They made the argument that an equilibrium **point** does not seem to be supported empirically, and that equilibrium orbits are more in evidence. The analysis is clearly harder in the latter case.

Parameter Extraction from a Nonlinear MOSFET Model

Client: Jet Propulsion Laboratory

The Clinic has been working for JPL on various aspects of MOSFET (metal-oxide-semiconductor field-effect transistor) modelling for four years. The equations governing current flow in a MOSFET (three nonlinear Poisson equations in the steady-state case) are too complicated to yield explicit analytical formulae except in idealized geometries, essentially the infinite-length, infinite-width case. Yet these formulae have worked very well in the past, especially when the parameters they contain are "extracted" or "identified" from test data. That is, the test data are used to extract a number of parameters (7 or 9, but sometimes more given the sophistication of the model), and then the formulae are available for use in circuit analysis over wide voltage ranges to good effect. Recently the pressure of miniaturization has forced re-evaluations of this process, including model extensions to finite-width, finite-length cases, and the experimental design considerations of the choice of width and length to use for testing. Additionally, the pressure of test and extraction costs has imposed design choices for reductions in the amount of test data collected. This most recent JPL clinic explored the nature of the models and the experimental design procedures with the goal of retaining good-fit parameter predictions from reduced data collection, [18].

Past experience has shown the choice of statistical model in the parameter extraction routine was not obvious. In particular, the most important questions under consideration were linear versus non-linear regression models and whether the error term in the model should be additive or not. In developing a proper statistical model, the team decreased the number of parameters to extract to four.

The theory for optimal (in some sense) design of experiments is rather well developed when based upon a linear model. Various attempts to linearize the MOSFET model were made but none worked satisfactorily. This made it necessary to work with a nonlinear model, the ratio between two polynomials of first degree. Existing statistical software packages were used to test for convergence, sensitivity in starting values, and overall performance. In order to take advantage of the behavior of the function under consideration, the clinic developed a computer program for the extraction of the parameters based on the Levenberg-Marquardt technique. Datasets were generated and different fractions of the datasets were used in order to illustrate the minimum size which would allow sound parameter extraction. This was done completely with generated data, which limits the possibility to draw general conclusions.

The time available for the clinic did not allow a deeper exploration of optimal designs for nonlinear regression models, a topic not well explored. However, by using a one step linearization of the model, one can get a fairly good idea about the optimal design in the nonlinear model, i.e., when the linearization is valid. Applying this optimal design theory to our particular problem involves substantial generation of computer code and it was not possible for the clinic to look any further into that area.

3.1 Acknowledgement

I am grateful for discussions with Professors Gunnar Andersson, Bob Borrelli and Stavros Busenberg during the preparation of this paper.

3.2 References

[1] SPANIER, J., *The Mathematics Clinic: An Innovative Approach to Realism within an Academic Environment,* Amer. Math. Mo. Vol. 83, No. 10, 1976, pp.771-775.

[2] SPANIER, J. AND BORRELLI, R. L., *The Mathematics Clinic: A Review of its First Decade,* UMAP Jour. Vol. IV., No. 1, 1983.

[3] CUMBERBATCH, E., *The Claremont Colleges' Mathematics Clinic,* **Teaching and Applying Mathematical Modelling**, ed. J. S. Berry et al., Ellis Horwood Ltd., 1984.

[4] GLASSFORD, A.P.M. AND GARRETT, J.W., *Characterization of contaminant generation characteristics of satellite materials; Phase II*

- *test method development, Interim Technical Report* Technical Report, Lockheed Missiles and Space Co., Inc., Dec. 1985.

[5] GLASSFORD, A.P.M. AND LIU, C.K., *Characterization of contaminant generation characteristics of satellite materials, Vol. I,* Technical Report, Lockheed Missiles and Space Co., Inc., Nov. 1983.

[6] GLASSFORD, A.P.M. AND LIU, C.K., *Characterization of contaminant generation characteristics of satellite materials, Vol. II,* Technical Report, Lockheed Missiles and Space Co., Inc., July 1984.

[7] SCIALDONE, J.J., *Characterization of the outgassing of spacecraft materials,* Technical Report SPIE 287, Shuttle Optical Environment, 1981.

[8] SCIALDONE, J.J., *Self contamination and environment of an orbiting spacecraft,* Technical Report TN D-6645, NASA, 1972.

[9] ZEINER, E.A., *Measurements of the kinetics and transport properties of contaminant related from polymeric sources in space and the effects on collecting surfaces,* Technical Report, Aerojet Electrosystems Co., 1976.

[10] FANG, W., EPSTEIN, E., GRAGG, L., LY, C., McNIEL, J., ZARON, E., SHILLOR, M., *Outgassing and Contamination in Vacuum Systems* Claremont Math Clinic, Final Report, 1988.

[11] BUSENBERG, S., FANG, W., SHILLOR, M., *Analysis of an Outgassing Contamination Problem,* in preparation.

[12] NEWLIN, L., FUTERMAN, J., GEE, B., MEISNER, V., PILORZ, S., TERZEMINELO, D., VENCIL, J., CHAPMAN, R., *Optimization of the Use of Mass Spectral Data,* Claremont Math Clinic, Final Report, 1985.

[13] TURNER,G., FALCONER, R., LITTLE, K., PEREZ, G., SANDOR, S., SARWAS, R., WRIGHT, D., HEIJMANS, K., *Computational Methods for the Mass Spectrometer,* Claremont Math Clinic, Final Report, 1986.

[14] CHAPMAN, G.R., *Optimal Data in Mass Spectrometry,* Math. Engng. Ind., Vol 1, No. 3, pp. 235-245, 1987.

[15] ORCUTT, F.K. AND NG, C.W., *Steady-State and Dynamic Properties of the Floating-Ring Journal Bearing,* J. of Lub. Tech., Jan. 1968, pp. 243-253.

[16] LI, C.H., AND ROHDE, S.M., *On the Steady State and Dynamic Performance Characteristics of Floating Ring Bearings,* J. of Lub. Tech, July 1981, Vol 103, pp. 389-397.

[17] HAMILL, M., FRANKEL, P., LIN, S., BORELLI, R., *Development of Mathematical Models of Floating Ring Bearings for High-Speed Turbo Machinery Applications to Analyze Bearing Stability,* Claremont Math Clinic, Final Report, 1986.

[18] ALLEN, D., FLEISHMAN, R., HAMZA, H., LACEY, S., LARSSON, K., PANAGIOTACOPULOS, D., VELASCO - HERNANDEZ, J., ANDERSSON, G., *Parameter Extraction from a Nonlinear MOSFET Model,* Claremont Math Clinic, Final Report, 1988.

4. Pattern recognition and a mathematical model for a railroad company. A project of the Waterloo Math. Clinic

by *B. Forte*

University of Waterloo, Waterloo Ont.

Canada

4.1 Introduction

The Math. clinic for industrial problems was set up at the University of Waterloo at the end of 1986. The University of Waterloo had already strong ties with industries, financial companies and government agencies through its co-operative educational programs. In the co-op programs students alternate four months of studies in one of the 5 Faculties with a work-term of four months in one of the almost 3,000 organizations employing.

In the Math. Clinic students are organized in teams of three and supervised by a faculty advisor. They work on the project during the summer months (May-August). The size of the project has to take into account this time limitation so that the team can complete the assignment by the end of August. Working for the clinic can replace the work-term in a co-operative program. While the Math. Clinic is run by the Department of Applied Mathematics, it is open to all the interested students in the 4 Departments of the Faculty of Mathematics.

There have been projects in industrial mathematics that were more involved and more demanding (concerning background preparation) than those considered suitable for the math. clinic.

These projects have been the subject matter of Master's and M.Phil. theses. As a sample, I present here in a outline, one project carried out by a Master student in a two year work for a steel company

(STELCO) regarding a particular process at its mill in Hamilton (Ontario-Canada): the batch annealing.

I shall also give the outline of a second project for the same company that has been carried out by the math. clinic in the summer 1987. It concerns another process at the mill: the cold rolling of steel.

Finally I will describe in more detail a recent mathematical model that has been developed by the clinic in the summer of 1988 for CP-Rail (Canadian Pacific Railway), it deals with: the evaluation of the rail head loss.

In all these projects I have been the faculty advisor.

4.2 The Batch Annealing of Steel Rolls

The batch annealing process at a mill consists of heating a stack of a certain number of steel coils and allowing them to cool slowly.

Each coil has been formed by tightly rolling thin steel sheets.

The purpose is to toughen the steel, to restore its original internal grain structure and improve its ductibility and flexibility. The coils have an average height of 100.00 cm., an average inner radius of 25.00 cm., an average outer radius of 80.00 cm. During the rolling of the steel sheets, gaps may have been produced between steel layers in which gas and oil and dirt may be trapped. The result is that the rate of the radial heat flow is not the same as that of the tangential and the vertical flows (which depend only on the conductivity of the steel).

Since each coil consists of more than 140 spirals, the mathematical model for the heat transfer through the coil would lead to an unwieldy system of equations and boundary conditions.

A first simulation is introduced by considering each coil as made up of thin concentric cylinders separated by even thinner layers of different conducting material. A correction in the radial conductivity is then introduced to simulate the heat transfer by conduction around the roll from one steel layer to the next.

Still one will be dealing with more than 4 x 140 equations to represent the heat transfer by conduction through each steel layer, the boundary condition between the steel layer and the material filling the gap, the conduction through this material, the boundary condition between the filling material and the next steel layer.

Another simulation is introduced. In this simulation the coil is considered as an inhomogeneous, anisotropic solid bounded by two coaxial cylinders. In this manner only one equation is necessary to describe the heat transfer in the coil. Cylindrical coordinates were adopted, thus the equation reads

$$\frac{\partial}{\partial t}(\rho c T) = \frac{\partial}{\partial} r(k_r \frac{\partial T}{\partial r}) + \frac{k_\theta}{r^2} \frac{\partial^2 T}{\partial \theta^2} + k_z \frac{\partial^2 T}{\partial z^2}$$

where r, θ, z are the cylindrical coordinates, T is the (absolute) temperature (unknown function of position and time), ρc is the steel specific heat, a function of temperature, k_r, k_θ, k_z are the thermal conductivities in each specified direction, also dependent on temperature and possibly position.

An improved Stikker's type formula for the radial conductivity k_r was introduced.

For the annealing process, the rolls are piled up in stacks of 3 to 4 rolls. Each is then capped and sealed inside the cap (the inner cover) to allow a special mix of gas to flow inside between the outer surface of the coils and the inner cover, between the coils (through separator plates) and inside the coils along their inner surface. The stakes (usually in the number of four) with their caps are enclosed in a furnace. However, it is the cap itself that in the model has been considered as the heat source for each stack. The timetable for the on-off switching of the furnace has been simulated by a special time dependent temperature distribution on the inner cover.

Even when one takes into account the whole system (inner cover, gas, coils, separator plates) the assumption of axial symmetry seemed to be reasonable and this reduced the mathematical problem to one in two dimensions. However the heat exchange of the outer surface of each coil with the circulating gas and the cover was taken into account. It was also taken into account the heat transfer from the gas to the coils through their inner surface. The gas is indeed the only heat source for the inner surfaces.

The heat transfer from the cover to the coils is by radiation, the heat transfer coil-gas is by radiation and convection.

A simplified model for the gas flow was introduced together with some parameters simulating the possibility of regulating the gas flow through the separator plates and along the inner surface of each coil.

The D.W. Peaceman and H.H. Rachford finite difference method has been used to solve the corresponding nonlinear partial differential problem. The method (an alternating-direction implicit procedure) has been used with a suitable non uniform grid and variable time step, to provide the necessary numerical stability and the required accuracy (within 2 %).

The computer program was developed on an IBM 4341 at the University of Waterloo. The program was written in IBM-extended Fortran H and employed routines from the IMLS subroutine library.

The model is flexible enough to allow:

a) a variable number of coils in each stack (1 to 5),

b) the absence or presence of separator plates,

c) different sizes and shapes of coils and separator plates,

d) changes in the geometry of the inner cover.

With the model one can conduct experiments on the effect of

a) the use of gas with different composition,

b) the adoption of different schedules for on-off switching of the furnace,

c) the choice of different gas flow,

d) the introduction of additional heat sources.

Of course, the objective of experimenting with the model is to reduce the anneal time while preserving and eventually improving the quality of the steel. A long term objective is the creation of a file of data for automatic control of the annealing process.

4.3 Cold Rolling of Steel

The aim of the cold rolling of steel is to reduce the gauge of the steel sheet. This is typically done by rolling the steel through a five stand tandem mill consisting of five couples of rolls that rotate in opposite directions. The opening in the two rolls in each couple can be adjusted by the mill operator. Each roll is about 10″ in diameter and typical initial gauge of a 0.12 percent carbon steel is 0.80″ to be reduced to a gauge of the order of 1/10 of that amount. The intent of this clinic project was to produce a mathematical model of the cold rolling of steel at a five-stand tandem mill for a computer program (in BASIC) to simulate the process. The computer will take the following input data:

1. initial gauge of steel,
2. final gauge of steel,
3. width of the steel strip,
4. carbon content of strip,
5. default values according to the operator's suggested reduction scheme.

The following limitation in the mill operation are to be taken into account:

1. maximum roll speed,
2. maximum load at each stand,
3. maximum strip exit temperature.

The model, to be effective, within 5 minutes on an IBM-PC should return

1. suggested reduction at each stand,

2. suggested interstand gauges,

3. mill speeds,

4. predicted strip exit temperature,

5. warning messages to the operator of the possibility of overheating the the strip and/or excessive power consumption.

The model had to take into account the following phenomena that take place in the process:

a) the slipping friction between the rolls and the strip,

b) the elastic compression of the strip,

c) the elastic deformation of the rolls,

d) the plastic deformation of the strip,

e) the elastic recovery,

f) the heat generated in producing the plastic deformation,

g) the frictional heat in the slipping of the rolls over the strip,

h) the heat transfer by convection from the strip to the surrounding,

i) the heat transfer by conduction from the strip to the rolls and through the rolls.

While other phenomena were considered negligible, the presence of all the above phenomena have been proven relevant for a realistic model.

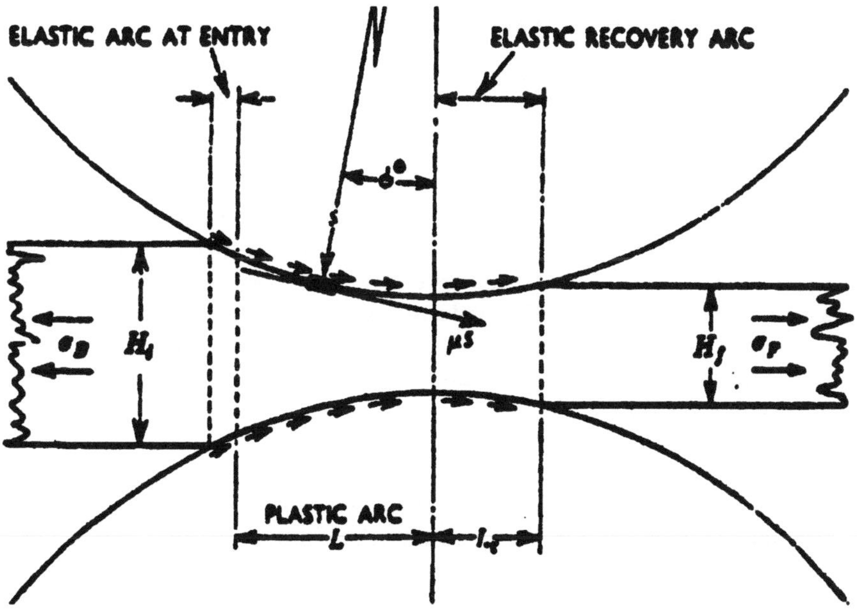

Figure 1.

In Figure 1, we show the various regions of the strip affected by the elastic and plastic deformations.

The algorithm starts with initial (default) values and uses an iterative procedure to converge, hopefully, to the best mill set-up. Experiments on the model have shown that if the initial data are far out the algorithm diverges. One does not yet know how close to the right values the initial values have to be in order to guarantee the convergence of the algorithm. The check for convergence is based on the sequence of calculated roll radii.

A first guess is made on the location of the so called neutral plane, which is uniquely defined by those points where sticking friction occurs, that is velocity of the strip is equal to the velocity of the rolls. The neutral plane is taken as plane of reference for all calculations and in the iteration changes position from cycle to cycle.

As it is at present the model can be used as reference by the mill operator. When fully tested and improved, it may be used to fully automate this phase of the mill operation.

4.4 Evaluation of Rail Head Loss

Using a wagon equipped with an optical imaging system, known as "Liteslice" the CP-Rail Company is able to pick up about 200 points of a cross section of the rail, while the train travels at about 60 mph. The acquisition of these data in the form of 200 x-y digitized coordinates which describes the worn rail shape is done at a rate of 5 images/sec. This means that the necessary information concerning each cross section can be fed into the on board computer at a rate of a cross section every 5-6 yards of rail. The objective of this project is to develop a computer program capable of determining in real time the extent of rail wear and find out whether or not it is in the "rail wear limits".

A first problem inherent to the "Liteslice" system had to be solved. The way the cameras (two per rail) are mounted above the running surface of the tracks to produce the image of the cross section results in a shadow underneath the head of the rail, missing a good source of information (see Figure 2).

Indeed, the knowledge of the part of the rail in the shadow would help in solving a second problem. This second problem consists in identifying the type of rail. CP-Rail has used 6 different types of rails (the 100 RE, 100 RE-HF, 115 RE, 130 RE-HF, 132 RE, 136 RE, the number indicates the weight in lbs/yrd for each type) and no accurate records of track placement has been kept.

The task of recognizing the type of rail we are dealing with from its present shape is made more difficult by the fact that of the 6 types of rail, there are three pairs with almost identical central zone (the web) and base.

To minimize the uncertainty about the rail type, its cross section has been divided into 4 zones easy to identify. From bottom to top (see Figure 2): the base, the fillet, the web, the head.

Both fillet and web are circles.

The cross section of the original rail is symmetrical with respect to a vertical line which is called "the centerline".

The first task in recognizing the rail type is to find from the data, this centerline. The data points between the bottom and the top of the web on either side were paired by minimizing the difference in their vertical (y-)coordinates.

A linear regression is performed on the midpoints of the pairs. The resulting line is assumed to be the centerline. It has been found that a difference of 1/4″ on the y-coordinates of the paired points gives at most a 1/100″ error in the horizontal position of the midpoint.

The base of the rail would be an excellent source of information to identify the railtype. The presence of stones, dirt, bolts, etc. make the data points on the bottom unreliable.

The most informative and sufficiently reliable zone is the fillet. The unique slope of the fillet at its midpoint and its distance from the y-axis were used to locate such point. The data points in the fillet zone were used to estimate the radius and the x-coordinate of the centre of the fillet. A modified least square method on nine data points in the fillet zone was devised to locate the fillet centre, having the radius and the x-coordinate of the fillet centre constrained to their expected values.

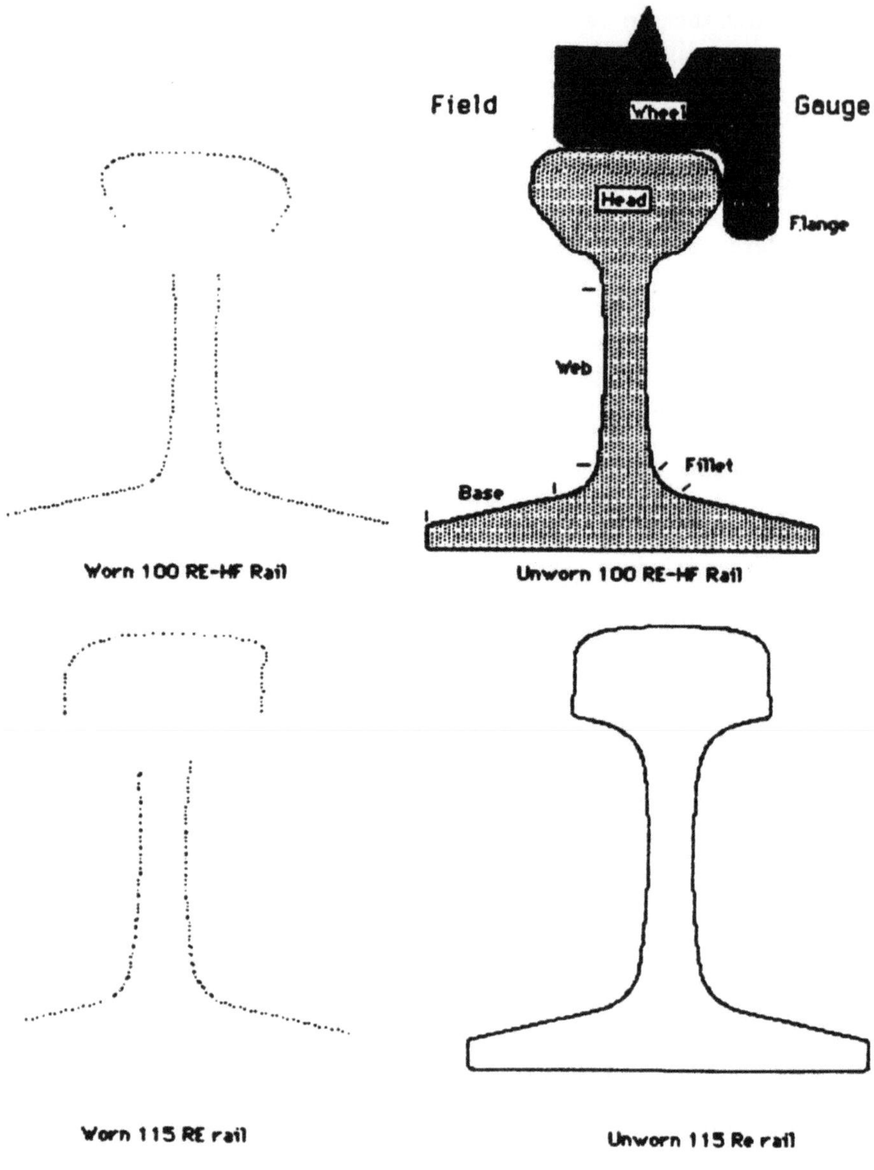
Field
Wheel
Gauge
Head
Flange
Web
Fillet
Base
Worn 100 RE-HF Rail
Unworn 100 RE-HF Rail
Worn 115 RE rail
Unworn 115 Re rail

Figure 2.

The (minimum) average distance of the points on the calculated fillet in each rail type provides an error parameter for the identification of the rail. Since there are two fillets, one for each side of the rail, one gets in this manner two error parameters per rail type. Two more error parameters are provided by the webs (one for each side). The radii of the webs are known for each rail-type. The location of the centre relative to the centre of the fillet is also known for each rail type.

A fifth error parameter is given by the minimal width of the web as derived from the data points and the known minimal width for each rail-type.

Using the error parameters, a special weighting scheme has been employed to find the rail type that most likely is the one under consideration.

At this point, knowing the rail type, one examines the wear of the rail head.

Typically there are two kinds of wear: a vertical wear and a horizontal one. There is loss of material on the top due to the contact between the wheel and the rail head, the weight of the train, the slipping friction etc. There is loss of material on the side due to the contact between the wheel flange and the rail. This is more relevant on curves. Also plastic flow takes place due to the weight of the train, which creates some kind of lips at the edges of the head. When a certain percentage of the rail head has been worn away, the rail must be replaced. Now, this percentage area has been translated into vertical and horizontal loss. To measure them, a parabola is fitted to a certain number of data points near the top of the head. The intersection of this parabola with the centerline is considered the top of the head of the worn rail the vertical distance of this point from the corresponding point in the unworn rail is taken as the measure of the vertical loss.

The horizontal line, a certain distance below the "top" of the worn head, is then considered. Then the midpoint of the two closest data points on either side of this horizontal line is determined. The horizontal distance of this point from the corresponding point (with the same y-coordinate) on the unworn rail measures the horizontal loss.

It is expected that the rate of image rejection for insufficient data will be very high. Since more than one image of cross section for each rail is available, second order statistics can be used.

4.5 Acknowledgments

The model for the batch annealing has been designed by D. Mundie and it was the subject of a thesis for her Master's Degree in Applied Mathematics (1981). The model was later improved and extended (to include the cooling process) by W. Hughes, he was at that time a research assistant on grant A-6199 by NSERC. The Math. Clinic team of undergraduate students that implemented the model for the cold rolling consisted of S. Butler, J. Riegert and B. Sutherland.

Finally the model of the rail head wear was the work of the Math. Clinic team consisting of F. D'Ippolito, V. Kolbas, R. Muir.

To all of them goes my own appreciation and that of the Department of Applied Mathematics of the University of Waterloo for work well done. Special thanks are due to the two companies, STELCO and CP-Rail, that provided the projects.

4.6 References

Batch annealing:

[1] Chang, Y.P. and Poo, K.C., "Heat Conduction in Cylinders of an Anisotropic Medium", Letters in Heat and Mass Transfer, vol. 6, pp. 293-299, 1979.

[2] Harvey, G.F., "Mathematical Simulation of Tight Coil Annealing ", The Journal of the Australian Institute of Metals, vol. 22, pp. 28-37, 1977.

[3] Kojima, H., Miki, T., Miyata, M and Tsuru, H., "Analysis and Application of Heat Transfer in Tight-coil Annealing Furnace", Nippon Steel Technical Report No. 12, Dec. 1978.

[4] Mundie, D. "A mathematical model of the batch annealing process", Master's Thesis, University of Waterloo, Dept. of Applied Mathematics, 1981.

[5] Stikker, U.O., "Numerical Simulation of the Coil Annealing Process", Mathematical Models in Metallurgical Process Development, 151 Pub. P123, The Iron and Steel Institute (British), pp. 104-113, 1970.

Cold Rolling:

[1] Bryant, G., "Automation of Tandem Mills", The Iron and Steel Institute, 1973.

[2] Devadas, C. and Samasekera, I.V., "Heat Transfer during Hot Rolling of Steel Strips", Iron Making and Steel Making, 13(6), pp. 311-321, 1986.

[3] Roberts, W., "Cold Rolling of Steel", Marcel Dekker Inc., 1978.

[4] Whitton, P. and Ford, H., "Surface Friction and Lubrification in Cold Strip Rolling", In Research on the Rolling of Strip, a Symposium on Selected Papers, 1948-1958, the British Iron and Steel Research Association, London, England, pp. 161-173, 1958.

[5] Wusatowski, Z., "Fundamentals of Rolling", Pergamon Press, 1978.

Evaluation of rail head loss:
No official literature available at the present time.

5. Industrial Mathematics & a Contest in Modeling

B.A. Fusaro

Department of Mathematical Sciences

Salisbury State University

Salisbury, Maryland, U.S.A.

Professor Capasso deserves credit for including a presentation on applied mathematics education in SASIAM's International Workshop on Mathematical Methods for Industrial Problems. The Mathematical Contest in Modeling (MCM), which will hold its fifth competition in February 1989, has benefited from the contributions of applied and industrial mathematicians. They help the MCM by providing problems and by serving as judges. On the other hand, participants in this workshop stand to gain from the MCM and similar education activities that increase the student's interest in applied mathematics. Invited speakers E. Cumberbatch, I. Stakgold and others have noted the scarcity of graduate students. E.A. Connors of the University of Massachusetts has said that (American) PhD mathematicians are becoming as endangered as the bald eagle. This shortage stems directly from a lack of college (undergraduate) students with an interest in applied mathematics. As the record will show, the MCM has elicited significantly more responses for each year of its existence.

The outline of the presentation is:

1. Background and Concept
2. Implementation (Unsuccessful)
3. Implementation (Successful)
4. Goals
5. The Record
6. Assessment

5.1 Background and Concept

The concept of a national applied math contest for undergraduates occurred to me in October 1983. It surfaced because of difficulties we were having in getting our students to prepare for the Putnam Mathematical Competition, an American examination for college students (undergraduates). The prestigious, venerable 50-year old Putnam has served as a screen for mathematicians. Salisbury State has a high percentage of first-generation students, who tend to be fearful about facing such a formidable examination alone. The Putnam practice of reporting very low scores adds to the chilling effect. Finally, the small applied content of the Putnam problems does not generate much enthusiasm in practical-minded students. Also, it would be difficult to tell from the Putnam, or any of the other mathematics competitions throughout the world, that the computer existed.

The concept was based on my experiences from the 1960's. I had chafed at the overemphasis in established mathematics of the pure, formalistic approach, almost devoid of content. Many campuses lacked any appreciable applied or constructive presence. In my mind, (classical) applied mathematics, computational mathematics, and statistics were as important a part of mathematical activities and curricula as pure mathematics.

These background thoughts merged and then popped up in verbal form as "Applied Putnam". My other colleagues liked the idea, but it seemed prudent to check with some off-campus mathematicians who had long involvements in applied math. Calls to M.S. Klamkin, H.O. Pollak, and E.Y. Rodin elicited favorable responses and encouragement to proceed.

5.2 Implementation (Unsuccessful)

Being Chairman of the Education Committee of the Society for Industrial and Applied Mathematics (SIAM) gave me a natural forum for this project. An outline of a proposal for a pilot contest was sent to members of this SIAM committee on 5 November 1983. The gist of the proposal is best presented in table form:

	Putnam	Applied Putnam
Time of Contest	December	March
Sessions	Two (each 3 hours long)	Two (each 3 hours long)
Number of problems	Twelve	Two
Type of Problems	Structural, Pure	Contextual, applied
Format	Individual Students	Teams of three students
Equipment	Paper & Pencil	Microcomputers

The committee liked the proposal, but with strong reservations on the time allotted per problem. The feeling was that an applied math problem could not be done in half a day - estimates ran from a day to a week. One experienced SIAM officer said that a realistic problem needed a whole semester! These observations, coupled with my fairly unshakable view that a contest for undergraduates should not occupy more than a weekend, doomed one of my favorite schemes: Teams should do one continuous and one discrete problem.

Although the Education Committee looked on the applied Putnam with favor, SIAM's Leadership felt that the committee already had enough projects and that it should continue to concentrate on other activities. However, so many people had judged the idea to be good and workable that I decided to seek another forum.

5.3 Implementation (Successful)

Warren Page, editor of the **College Mathematics Journal**, gave an invited lecture to our Mathematical Association of America section on 18 November 1983 in Washington, D.C. His lively presentation included many applied examples, so I approached him after his talk. His initial reaction was that the concept was interesting but not feasible. He didn't seem to like the term "Applied Putnam".

About three weeks later, Page called me at home to say he had given this concept of an applied contest quite a bit of thought. He had come around to the position that it was a valuable idea and that it ought to be done. Moreover, he had talked to S.A. Garfunkel,

Executive Director of COMAP. The Consortium for Mathematics and its Applications had been supporting applicable mathematics in a variety of ways since 1972.

Garfunkel was very enthusiastic, and Page urged me to get in touch with him. Although I was a member of COMAP and had used its materials, I had never had any interaction with Garfunkel. After one phone conversation, it was clear that we had similar goals. In fact, my personal campaign to "increase the applied presence on campus" might be one way to describe what COMAP had been doing over the years. He suggested that a proposal for a three-year grant be sent to the U.S. Department of Education, with COMAP administering body and me as project director. The offending term "Applied Putnam" was transformed into "Mathematical Contest in Modeling". A preliminary proposal squeaked by the 12 January 1984 deadline and the final three-year proposal was approved in June 1984.

5.4 Goals of the Mathematical Contest in Modeling (MCM)

The goals and purposes of the MCM are best described by two paragraphs from the abstract of the final proposal:

- The selection of realistic open-ended problems chosen with the advice of working mathematicians in industry and government.
- An extended period for teams to prepare solution papers within a clearly defined format.
- The ability of participants to draw on outside resources including computers and texts.
- An emphasis on clarity of exposition in determining final awards with the best papers published in professional mathematics journals.

As the contest becomes established in the mathematics community, new courses, workshops, and seminars will be developed to help students and faculty gain increased experience with mathematical modeling.

Garfunkel and I were in firm agreement that the contest must be primarily an educational experience, and not a competitive one. In a sense, we wanted it to be closer to the spirit of traditional English sports rather than to modern American sports.

The following two problems will provide the flavor of the MCM. Each team had to choose one of the problems Friday morning and complete work on it Monday afternoon.

5.4.1 Managing an Animal Population (1985)

Naturally-occurring animal populations are found in environments where there are resource constraints, that is, limited food, space, water, etc. Select a fish or mammal (Chincoteague ponies, deer, rabbits, salmon, striped bass, for instance) and an environment for which you can find appropriate data, and formulate an optimal policy for harvesting members of the population.

5.4.2 Managing a Strategic Reserve (1985)

Cobalt, which is not produced in the U.S., is essential to a number of industries (Defense accounted for 17 percent of the cobalt production in 1979). Most cobalt comes from central Africa, a politically unstable region. The Strategic and Critical Materials Stockpiling Act of 1946 requires a cobalt reserve that will carry the U.S. through a three-year war. The government built up a stockpile in the 1950's sold most of it off in the early 1970's, and then decided to build it up again in the later 1970's, with a stockpile goal of 85.4 million pounds. About half of this stockpile had been acquired by 1982.

Build a mathematical model for managing a stockpile of the strategic metal cobalt. You will need to consider such questions as: How big should the stockpile be: At what rate should it be acquired: What is a reasonable price to pay for the metal? You will also want to consider such questions as: At what point should the stockpile be drawn down? At what rate should it be drawn down? What is a reasonable price to sell the metal? How should it be allocated? (Accompanied by a sheet with more information on the sources, cost, demand, and recycling aspects of cobalt.)

The ground rules allowed the team to use any inanimate source. It was emphasized that partial solutions were acceptable, and that there was no passing score. Except for diagrams, graphs, etc., the solution paper had to be typed.

The open-ended nature of the problems required guidance, so the following outline was provided:

1. A clarification or restatement of the problem as appropriate.
2. A clear exposition of all assumptions and hypotheses.
3. An analysis of the problem, justifying or motivating the model to be used.
4. The design of the model.
5. A discussion of how the model can be tested.
6. A discussion of the strengths and weaknesses of the model(s), including error analysis and such things as stability.
7. A one page abstract or summary.

5.5 The record

We expected about 50 teams the first year, but got 90. The contest has grown about 30% each year, as the following table shows:

	1985	1986	1987	1988	1989
Number of colleges with active teams	70	89	118	132	?
Number of teams that submitted solution papers	90	118	156	202	?
Number of publishable papers	6	6	4	6	?

Since each team consists of three students, these numbers reflect a sizable student participation.

Outstanding solution papers have been published in Pergamon's **Mathematical Modelling**, an international journal [1,2,3]. Outstanding teams have been invited to make presentations at national meetings of ORSA and SIAM.

5.6 Assessment

The nature of the contest makes assessment or judging very difficult. The papers are first run through "triage", a preliminary assessment. The judges spend about 7 minutes per paper and the assessment is based only on the abstract and overall structure. About 25% of the bottom papers are eliminated and the rest are sent for final grading.

The papers are then read again, spending about 10 minutes per paper, ranked, and approximately the bottom 20% eliminated. This process is done a total of three times. The reading time per paper is increased each time.

The target is to have about 20% of the papers Meritorious and 20% Honorable Mention. The remaining 45% are classified as Successful Participants. (Only one paper has been disqualified in the four years of the contest.) About six "winners" are then chosen from among the Meritorious papers. This is a very difficult time of the grading, and tempers often run high. The solution papers that are chosen as winners are published. Moreover, two or three winning teams will be designated as **ORSA** or **SIAM** winners. These teams are invited to make presentations at ORSA's and SIAM's national meetings. The teams also have their transportation paid and receive cash prizes. These undergraduates have always given excellent presentations. Readers are urged to examine their published papers [1,2,3] to see what a team of three college students can accomplish on a weekend.

5.7 References

[1] B.A. Fusaro (Editor), "Mathematical Competition in Modeling", **Math. Modelling** (Pergamon), Special Issue, Vol. 6, No. 6 (1985).

[2] Ibid

[3] B.A. Fusaro (Editor), "Mathematical Contest in Modeling", **Math. Modelling** (Pergamon), Special Issue, Vol 9, No. 10 (1987).

6. Remarks on Industrial Mathematics

J. L. Lions

College de France and CNES

6.1 Introduction

Let us consider the "system theory" approach, more precisely the theory of **distributed systems** - i.e. systems governed by Partial Differential Equations. (P.D.E.).

This theory rests on three main aspect:

(i) **the state equation** (here a P.D.E.!)
(ii) **the nature of the control variable**
(iii) **the criteria** (or cost function).

The criteria is a function of the control and of the state which is itself computed through the state equation.

The problem is then to **construct** a control which **satisfies the constraints** (physical or, and, economical) and which **minimizes** (at least locally and, hopefully, globally) the criteria.

Here we want to address briefly some **not completely usual points connected with these three aspects**, and to indicate **the industrial applications** of these remarks.

The first (and obvious) question connected with (i) is: **do we know the state equation**?

A variant of the question is: is the state equation we are using **reliable**? This is the problem of the **validation** of the state equation.
We emphasize the **fundamental** importance of this "validation" question, since state equations are taken more and more seriously and are **actually used** in more and more applications. Remarks on

these questions are presented in Section 6.2, “Is the state equation known?.”

The second point we want to address concerns the nature of the control variable. It can be **stochastic** - such as in problems of **management**, a question we do **not** consider here, referring for it to the lecture of M. ROBIN, these proceedings.

The control variable can also be a **geometrical domain**: we have then the problems of **optimum design** were industrial applications abound.

The third point (iii) concerns the **cost function**. It can be a quadratic criteria, a classical situation that we do not consider here. We rather consider here cost functions which are constructed **in connection with properties of the solution** (s): free boundary (or free surface), turning points etc. Industrial applications are indicated.

We then consider in Section 6.5, “Flexible structures,” the case of **complex structures** where applications should become more and more important in the near future.

We do **not** consider here the application of the ideas of optimal control theory in numerical analysis nor the minimization algorithms which have an industrial interest of their own.

6.2 Is the state equation known?

Let us consider the state equation given by

$$(1) \qquad \frac{\partial y}{\partial t} - \Delta y + \sum_{i=1}^{n} a_i(x) \frac{\partial y}{\partial x_i} = 0 \quad \text{in } \Omega \subset \mathbb{R}^n \,,\ t > 0$$

where the a_i's are **partially known**.

We dispose of **measures** on y:

$$(2) \qquad (y \,, \mu_j) = m_j \quad j = 1 \,, \dots \,, N$$

where μ_j are the "sensors".

Examples:

$$\mu_j = \delta(x - x_j)\,\delta(t - t_j) \Rightarrow (y\,,\mu_j) = y(x_j\,,t_j),$$

μ_j = averaging process near $x_j\,,t_j$, *etc.*

Remark 1

If $\mu_j = \delta(x - b_j)\,\delta(t)\,,\quad b_j \in \Omega\,,$ (2) gives

(3) $\quad y(b_j\,,0) = m_j\,.$

If b_j "spans" Ω, (3) corresponds to the **initial data**.
We also assume (for the time being) that we know the **boundary condition**, say

(4) $\quad y = 0$ on $\Gamma \times (0\,,T)\,.$

How can we compute the a_i's and the state y?

Let us think of the a_i's are control functions and let us also think of $y(x\,,0) = y(0)$ as a control function

(5) $\quad y(0) = y^0\,.$

Then y^0 and a_j given, we can compute $y = y(x,t;a_i, y^0) = y(a_i, y^0)$ and **we want that this solution** (the **state**) **satisfies** (2)(and (3)) **"in the best possible way"**.

In the exact spirit of GAUSS and of LEGENDRE;, we then introduce the **"cost function"**

(6) $$J(a_i\,,y^0) \doteq \sum_{j=1}^{N} \left[(y(a_j\,,y^0)\,,\mu_j) - m_j \right]^2$$

and we look for

(7) $$\inf_{a_i,\,y^0} J(a_i\,,y^0)\,.$$

In (7) we look for the inf. taking into account all the constraints we may already know on the coefficients a_i's and on y^0.

"The" solution of (7) gives "the" coefficients a_i and "the" initial condition we are looking for.

A number of Remarks are in order.

Remark 1

Of course the above presentation is an over-simplification of more realistic situations. But the ideas are those actually used in practice.
The method is completely general:

1. The same type of method can be used for hyperbolic problems. For example when the state equation is given by

$$\rho(x)\,\frac{\partial^2 y}{\partial t^2} - \Delta y = 0 \tag{8}$$

where $\rho(x)$ is only partially known: $0 < \rho_0 \leq \rho(x) \leq \rho_1$ in Ω.

2. We can consider as well **non linear** state equations.

3. One can adapt the above ideas to situations where boundary conditions are only partially known.

Remark 2

Applications arise in cases where access to the system can present difficulties - such as in the **geological questions** - with applications in the **petroleum industry**. cf. [1] and the Bibliography therein.

Remark 3

Access to the complete system is also difficult in **medial** applications. cf. [2] and the Bibliography therein.

Remark 4

Another situation can arise where we have a state equation given by, say

$$\Delta y = 0 \quad \text{in } \Omega \subset \mathbb{R}^n \tag{9}$$

with no boundary condition known on part Γ_1 of $\Gamma = \partial\Omega$ but with **2 boundary conditions on $\Gamma_0 = \Gamma/\Gamma_1$:**

$$(10) \qquad y = g_0 , \quad \frac{\partial y}{\partial v} = g , \quad \text{on } \Gamma_0 .$$

This is a **non well set** problem where one can use again ideas and techniques of optimal control theory. cf. [3] [4] [5].

Remark 5

Problems of the preceding type also enter the family of the so called **"Inverse problems"**. Problems of this type also arise in situations where one measures a **function of the state** and not the state itself.
This is a very general situation!
Examples abound - **For instance** it is quite the usual case for space observations (remote sensing).

Remark 6

Situations where one looks even for the **structure** of the equation also arise. For instance in **structural mechanics** what kind of "memory" should we use? cf. [6].

Remark 7

There is another way of looking at these questions: what can we compute which is somewhat **insensitive** to these parts of the system which are either unknown or out of reach. An **attempt** is made in [7] [8] - but numerical applications have **not** yet been made of these ideas.

Remark 8

A situation where the **initial data** is only partially known arises in a systematic fashion in meteorology (and in Oceanology as well). The problem of finding the "best" initial condition is then known as the "assimilation problem". cf. Ph. COURTIER [9] - and the Bibliography therein.

6.3 Optimum design

Let us **assume** now that we know the state equation and the boundary (and initial) conditions that go with it. We consider situations where one can **act** on the system - through a **control variable**.

If **the control variable** is **the domain itself**, it means we are dealing with problems **of optimum design**.

One has to find the **shape** of a body such as a given **"cost function"** is minimized.

Examples

Minimize the drag. Or find the "best" shape of the "noise" of a plane such as to minimize "turbulence around the wings". cf. [10] and the Bibliography therein. This is **a very important** question in Aeronautical industry. cf. also [11].

Remark 9

From a technical view point, the first question which arises is the **representation mode** of the unknown part of the boundary. It can be made by "local variations" around a surface of reference. It can also be made by using a "cinematic representation" of the unknown surface.

Remark 10

In practical applications, one has always to find the "best" shape starting from a **suspected** good shape. This avoids the difficulty of having to consider **topological changes** in Ω.

Remark 11

The problem of optimum design can in some cases be "reduced" to a more standard problem of distributed control. For instance in the optimization of the shape of a glass [12] the problem amounts to find the best strategy of heating a furnace such as to achieve a given repartition of temperature.

Remark 12

For general references on the questions of optimum design, cf. [13] [14] and the Bibliography therein.

6.4 The cost function

We want now to give two examples of "non standard" cost functions, coming from industrial applications.

The first one concerns steel industry, and more precisely the cooling of metal in fusion.

The state equation is given by the **heat equation** with **two phases**: the liquid and the solid one. (This is again quite a simplification!). The two phases are separated by a so called **free boundary** (the interface).

The problem is to steer the metal in fusion to the solid state in minimum time and in the most homogeneous way. cf. [15], i.e. to derive the free boundary out of the plane where the metal goes out to be rolled up.

The **control function** consists in the possible policies of sending proper amounts of cool water at proper regions and during proper times.

Technically one has a non linear state equation. cf. [15] and the Bibliography therein.

In problems arising, among other things, in Chemistry and Biochemistry (cf. [16] and the Bibliography therein) one meets systems which have **several states solutions** (multi-state systems). This situation also arises in Mechanical structures (Buckling) and is also referred to as "bifurcation problems" . Other examples arise in, for instance, sodium heated water evaporation [17].

The cost function is defined in terms of **particular** solutions of the state equation which has, in that case, several solutions (and which may even have an infinite number of solutions). It can be the "minimum" solution, it can be some topological properties associated with the solution. Possible techniques are either global approaches [18] or continuation methods of the type of [19] (which gives a very useful code!).

6.5 Flexible structures

Improvements in the modelling and in the design of structures lead to the possibility of envisioning lighter (and flexible) structures.
The interest in aerospace industry is obvious. It is also of economical interest (both in space and on the ground!) to devise lighter structures, often with **composite materials**[1] .

Very many (interesting) questions immediately arise.

The first one (as ever) is to properly model such a system, which can consist of **a very complicated set of structures**. Think of a space structure, of a marine structure or of a robot. These structures can consist in a (large) number of pieces which can be reasonably considered as an assembling of pieces which are 3 dimensional **or** 2 dimensional **or** one dimensional. Modelling of these structures is an interesting and difficult problem in itself. One can use a number of approximations techniques.
Asymptotic methods seem to be very promising [20].

The next question is concerned with **the validation** of the models cf. [6].

The following question is then concerned with the **stabilization** of such a structure. Several methods are now under development [21] [22] [23] [6].

[1] Composite materials are also an interesting example of "industrial" mathematics.

6.6 References

[1] G. CHAVENT and Al. Publications of INRIA, Paris.

[2] E. MAGENES and Al. Publications of the L.A.N. of Pavia.

[3] J.L. LIONS. **Sur le Contrôl Optimal des Systèmes Gouvernés par des Equations aux Dérivées Partielles**. Paris-Dunod, Gauthier-Villars, 1968.

[4] J.L. LIONS. **Some Methods In the Mathematical Analysis of Systems and their Control**. Science Press Bejing and Gordon Breach, New York; 1981.

[5] R. LATTES and J.L. LIONS. **Méthode de Quasi Réversibilité et Applications** Paris-Dunod 1967.

[6] T. BANKS. Lecture at INRIA; June 1988.

[7] J.L. LIONS. Sur les sentinelles des systèmes distribués. 2 notes C.R.A.S. Paris, 1988.

[8] J.L. LIONS. Sentinels for periodic distributed systems Chinese J. of Math . 1989. Sentinels for Distributed systems with incomplete boundary data. Workshop on Mathematical Ecology. Trieste ICTP. November 1988.

[9] Ph. COURTIER. Application du contrôle optimal à la prévision numérique en Météreologie. Thesis. Paris 1987.

[10] R. GLOWINSKI and Al. Publications of INRIA. Paris.

[11] E . JAMESON. Personal communication. July 1988.

[12] F. MEHL, P. VILLION, J.P. YVON. Optimization of industrial glass furnaces. Numerical Methods in thermal problems. Vol. IV. Swansea 1985.

[13] J. CEA and Al. Publications University of NICE.

[14] O. PIRONNEAU. **Optimum Design**. Springer. 1985.

[15] Ch. SAGUEZ. INRIA Report and SIMULOG Publications. Work in cooperation with the Steel Industry .

[16] J.P. KERNEVEZ. **Enzyme Mathematics**. North Holland. 1985

[17] G. JOLY and J.P. KERNEVEZ. Heat transfer Optimization in Industrial Systems with multiple steady states. IFIP Symposium. Saint Jacques of Compostelle July 1987.

[18] C.L. LIONS **Contrôle des systèmes distribués singuliers**. Paris Gauthier Villars. 1983.

[19] E.J. DOEDEL and J.F. KERNEVEZ. Auto: Software for continuation and bifurcation problems in Ordinary Differential Equations. Applied Math. Tech. Report. CALTEC 1986 - (226 p.)

[20] Ph. CIARLET. Works and books in asymptotic methods in Structures Cf. in particular North Holland, 1988. Cf. also in publications of University Paris VI, LE DRET and the Bibliography therein and Collection RMA, Masson.

[21] R. GLOWVINSKI, C. LI, J.L. LIONS, Japanese J. of Applied Math. 1989.

[22] J.L. LIONS. **Contrôlabilité exacte, Perturbations et stabilisation de systémes distribués**. t.1. Contrôlabilité exacte. t.2. Perturbation Masson, RMA, 1988.

[23] J. LAGNESE and J.L. LIONS. **Modelling, Analysis and Control of thin plates**. - Masson R.M.A. 1988.

7. Collaborating with Industry - a Mathematician's Viewpoint

Sean McKee

Department of Mathematics

University of Strathclyde

Glasgow

7.1 Introduction

In October 1979 I had the task of setting up a consortium, the University Consortium for Industrial Numerical Analysis (UCINA), from scratch. This was originally a group of numerical analysts from 5 southern universities including Oxford where it was based. One day meetings were held two or three times a year between Academia and Industry and these led to an exchange of problems and information, which in turn occasionally led to contract work. When I left in January 1986 UCINA was self-funded with sufficient capital to pay my salary, my secretary's and travel expenses, and through individual contracts, to fund a number of research assistants.

Perhaps a measure of UCINA's success was that I was head-hunted by one of my clients, Unilever Research at Colworth in Bedfordshire. After some negotiation it was agreed that I should be paid (an industrial) salary and allowed to work half-time for the University of Strathclyde, a university 363 miles north and situated in the centre of an old, but now vibrant city by the name of Glasgow. In this talk I wish to concentrate on some of my research activities during this period of two and a half years.

7.2 Unilever Research

Unilever research is an Anglo-Dutch multinational with its core business based on oil. There are few domestic products that it does not manufacture: it makes ice-cream and ice-lollies which it sells under different brand names in different countries; the same is true of detergents, toothpaste, face creams, shampoos and hair conditioners and margarine; it is involved in the food industry, particularly the frozen food industry, and makes anything from meat paste to complete dinners; it owns palm and tree plantations around the world; in the U.S., Lipton's Teas is one of its companies; it is involved in the animal food's business and has recently successfully entered the medical products business. Another interest it has lies in fish farming. In short it is an extremely diverse company, with its research tending to be towards the biological sciences, rather than the physical sciences. In some ways this makes the mathematician's job extremely difficult, for be must be a generalist: there is no room for specialisation as there is, for instance, in the glass or electricity industries.

Unilever Research has three laboratories, Colworth, Port Sunlight and Vlardingen. Colworth is the largest, employing 1200 staff, most of whom are working scientists. It is at this laboratory that I am employed. In fact, when I first arrived I was given a room in the Georgian Manor House. Indeed, my room was the one with the open window, affording wonderful views over rolling lawns (Colworth has 1200 acres which include croquet lawns, a golf course etc..) However, on the arrival of the new managing director - senior staff appeared to turn round every nine months - he discovered this room for himself and I was ousted to the modern library block on the right of the picture.

This talk, for reasons of confidentiality, will deal superficially with a selection of different problems in which mathematicians can play an active and positive role. It will then proceed to discuss four industrial problems in slightly greater detail. Throughout this talk the objective will he to emphasise the need for the mathematician to be a generalist and be capable of working within an interdisciplinary team.

Toothpaste consists of silica and actives held together in a polymer gel. One of the major problems is the release of actives, particularly fluoride, to the plaque or tooth wall. Simple diffusion-type models can be constructed which give an indication of the relative rate of release and and uptake of different toothpastes with different rheological properties.

The question was posed, "How does one optimise the photosynthesis of palm trees (and can mathematics help in this direction)?" This is not untypical of the sort of question that a scientific industrialist might ask. This question is highly complex and not sufficiently precise and, as such, it is unlikely that mathematics could usefully contribute to its solution. However, mathematics can be applied to certain aspects of the problem. For instance, transpiration of an individual leaf can he modelled quantitatively, as can the amount of light falling on leaf surface.

A moist skin is important for good health and the maintenance of good looks, and the latter is the principal reason for skin creams. To what extent does humidity or lack of it affect the moisture (or, more precisely, the water activity) in the skin. This was studied by means of a discrete diffusion model - discrete because there are only a finite number of layers of skin.

Shampoo bottles, like Timotei, are now designed using solid modelling codes such as GEOMOD by an interdisciplinary team which includes a psychologist. The design is then interfaced with the stress analysis code MARC which produces a robust product with customer appeal.

Ice-cream consists of water, air, ice and oil and since oil is relatively expensive considerable effort is expended in determining how more air can he included in the product. The cost of manufacturing ice-cream is now comparatively small, but the cost of the underlying technology is not. One problem that Unilever overcame a number of years ago was shrinkage. When ice-cream is stored in a thermal gradient (i.e. the chilled cabinets one finds in supermarkets), it was found that after a period of about three weeks there was a sudden shrinkage.

Prawn and fish farming is a recent new business venture. In order to understand their reproductive cycle a study of their ecology was necessary; this work was started, and then stopped midway through for lack of funds.

This frustrating problem is typical and one that all industrial scientists experience from time to time. However, a similar ecological study was required for agribusiness in connection with the ecology of a parasite known by the generic term, coccidia. This will be discussed in a little more detail in 7.3.3, "Coccidia in Chickens."

Many medical products now depend upon an antibody-antigen reaction. A successful pregnancy testing kit called Clear Blue is one such, where the reaction takes place on the wall. In 7.3.1, "Reaction and diffusion at a cell wall"the mathematical problem is described in greater detail.

Two dimensional gel electrophoresis is an effective tool for the separation of protein macro-molecules. In order to determine if two samples are of the same genotype a comparison between two sets of spots is necessary. After filtering and spot identification and possibly the application of a rule-based system the problem of spot matching (albeit now of a smaller sample) still remains. This problem was identified as a generalisation of a matching of a bipartite graph. An algorithm was constructed and shown to be of polynomial type (Swailes and McKee (1988)). The complete project, including the expert system, was done under the Alvey programme; the partners involved were the University of Strathclyde, Plant Breeding Institute, Unilever Research and the Turing Institute.

7.3 Typical Problems in Industrial Mathematics

I have been connected with the following four problems to a lesser or greater extent. The first problem came to John Ockendon at Oxford from Unilever Research whilst I was UCINA Coordinator, and involves a reaction-diffusion problem. The second represents work done with the electronics company Plessey; this was completed with a student while holding my joint appointment. The third involves the theoretical ecology of a parasite that commonly occurs in the domestic chicken. This was completed when I was at Colworth by an interdisciplinary team of three mathematicians and three immunologists. Finally a rotor stacking problem will be mentioned. This arose from Mechanical Technology Ltd. when they presented it at the first Rensselaer Polytechnic Institute industrial study group. The work was completed by myself and a co-author (McKee and Reed).

7.3.1 Reaction and diffusion at a cell wall

Consider a simple reversible reaction between two reactants X and Y to produce a complex XY; in chemical notation

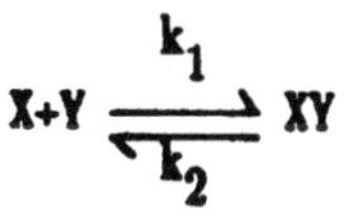

Figure 3.

where k_1, k_2 are the speeds of the forward and backward reactions, respectively.

In the systems of interest, one species, say Y, is immobilised on a side wall of the cell, and the other species X is dissolved in solution. The reaction takes place only on the side wall. The immobilised species Y is a specific binding agent for the dissolved species X; an example of such abinding agent is lectin concanavalin A which is specific to carbohydrates and glucose.

If a solution X is introduced at time t = 0 to the cell, then as the reaction at the wall proceeds, an X concentration gradient develops, and X diffuses to the wall until equilibrium results.

The aim of this study is to predict

(i) the X concentration profile u

(ii) the concentration θ of the complex XY at the reaction side wall as a function of time.

With the nondimensional variables $u = u(x,t)$ and $\theta = \theta(t)$ a one-dimensional model can be written down as:

$$(1a) \qquad \frac{\partial u}{\partial t}(x,t) = \frac{\partial^2 u}{\partial x^2}(x,t), \quad 0 < x < 1, \quad t > 0$$

$$(1b) \qquad \left.\frac{\partial u}{\partial x}(x,t)\right|_{x=0} = 0$$

(1c) $$\left.\frac{\partial u}{\partial x}(x,t)\right|_{x=1} =$$

$$= \frac{E}{1+L}\left\{(1-m)\,u(1,t) + L\left[1 - \int_0^1 u(x,t)dx\right] - u(1,t)\int_0^1 u(x,t)dx\right\}$$

(1d) $$u(x,0) = 1$$

Here the non-dimensional constants of the system are:

(a) the molar ratio $m = a_0 / c_0 d$ where a_0 is the initial Y concentration, c_0 is the initial X concentration at the reaction side wall, and d is the cell width.

(b) the reaction time scale ratio $L = k_2 / k_1 c_0$

(c) the diffusion reaction time scale ratio $E = (k_1 c_0 + k_2)d^2 / D$ where D is the diffusion coefficient of X.

In addition there is the conservation of the total number of species X, either in solution or bound in the complex XY. This is given by

(2) $$m\theta(t) + \int_0^1 u(x,t)dx = 1 \quad \text{for all } t \geq 0$$

By the standard use of Laplace transform techniques is is possible to reduce the problem, defined by (1) and (2), to the nonlinear Volterra integro-differential equation

(3) $$\frac{d\theta(t)}{dt} = \frac{E}{1+L}\left\{L\theta - (1-\theta)\left[1 - m\int_0^t \frac{d\theta(s)}{ds}k(t-s)ds\right]\right\}$$

where

$$k(t) = (\pi t)^{-\frac{1}{2}} \left(1 + 2\sum_{n=1}^{\infty} \exp(-n^2 / t)\right)$$

which then admits a solution through product integration techniques (see, e.g. McKee and Stokes (1983)). Further details can be found in Burgess et al (1986). The solution of this problem for various parameter values contributed to the calibration of several medical products, like Clear Blue.

7.3.2 The design of transducers for underwater vessels

It is necessary to relate the output of a pressure transducer to the incident sound field in order to he able to predict the performance of sonar arrays. The transducer has to be supported in some way and it is often convenient to mount it flush with the surface of a metal plate or baffle which is both finite and flexible.

A sound wave incident on a plate will give rise to transmission through the plate, reflection from it, and diffraction from the edges, so that a transducer mounted flush with the plate surface will be subject to the incident, reflected and diffracted pressure fields.

The tests were done by mounting the plate vertically under water in such a way that it can be rotated about the vertical axis. A sound projector is placed some distance away so that the sound field at the plate is virtually planar and the output of a central transducer is observed as a function of the angle of incidence of the sound wave for a range of frequencies from 50 Hz to 12 KHz. Figure 4 shows an example of a transducer at the centre of a plate.

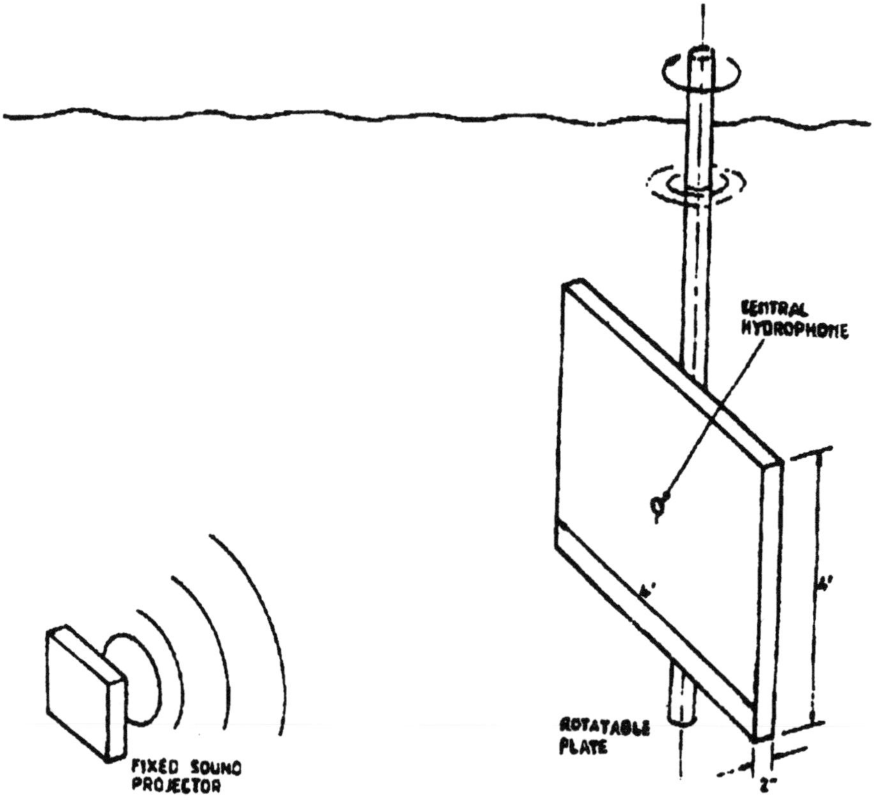

Figure 4.

The underwater pressure and the lateral deflection of a finite flexible plate are described by the equations

$$\nabla^2 p + (\omega / c)^2 p = 0 \quad \text{in } R - [-1,1]$$

$$B \frac{\partial^4 u}{\partial \xi^4} - \tau\omega^2 u(\xi) + p(\xi, 0^+) - p(\xi, 0^-) = 0, \qquad -l \le \xi \le l$$

together with the boundary conditions

$$\rho\omega^2 u(\xi) = p_\eta(\xi,0^\pm), \qquad -l \le \xi \le l$$

$$u''(\pm l) = u'''(\pm l) = 0$$

and

$$p(\xi,\eta) \sim e^{-i(\omega/c)(\xi \sin\theta + \zeta\cos\theta)} \text{ as } r^2 = \xi^2 + \zeta^2 \to \infty$$

Here $p(\xi,\zeta)e^{i\omega t}$ denotes the fluid pressure and $u(\xi)e^{i\omega t}$ the lateral deflection of the plate in the ζ – direction (see Figure 5).

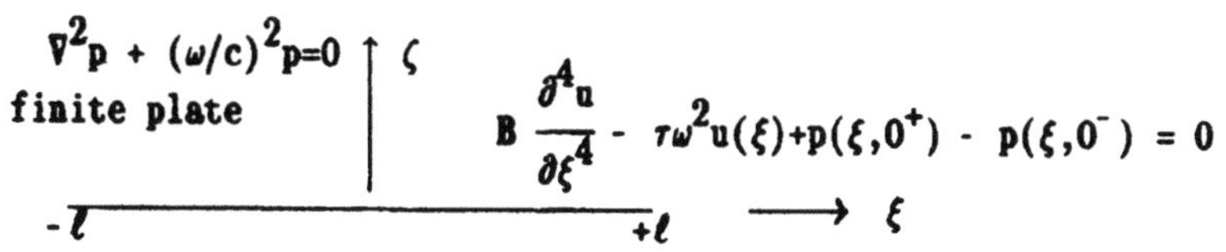

Figure 5.

Here the following notation has been employed:

c - sound speed in water

ω - frequency of incident wave

θ - angle of incidence of pressure wave

ρ - density of the plate

τ - mass of plate per unit area

B - bending stiffness of plate

$2l$ - width of plate

Subtracting the incident wave

$$q(\xi,\zeta) = p(\xi,\zeta) - e^{-i(\omega/c)(\xi \sin\theta + \zeta\cos\theta)}$$

and nondimensionalising the equations and boundary conditions using

$$x = \xi/l, \quad y = \zeta/l, \quad \alpha = l\omega/c, \quad S = \alpha \sin\theta,$$

$$C = \alpha\cos\theta, \quad \lambda = l\rho/\tau, \quad \beta^2 = B/(\tau l^2 c^2),$$

$$\psi = iq/c, \quad u = i\tau c^2 u/(l^2 c)$$

results in

(4a) $$\nabla^2 Q + \alpha Q = 0 \text{ in } R^2 - [-1,1]$$

(4b) $$\beta^2 \frac{d^4 U}{dx^4}(x) - \alpha^2 U(x) + 2Q(x,0^+) = 0$$

together with

(4c) $$\frac{dQ}{dy}(x,0^{\pm}) = \lambda\alpha^2 U(x) - e^{iSx}, \qquad -1 \le x \le 1$$

(4d) $$U''(\pm 1) = U'''(\pm 1) = 0$$

(4e) $$Q(x,y) \to 0 \text{ as } x^2 + y^2 \to \infty$$

The general approach employed was to assume that the flexible plate was considerably longer than it was wide. Green's functions of equations (4a) and (4b) were then sought and coupled through (4c) resulting, after some nontrivial analysis, in a Cauchy integral equation of the form

(5) $$\frac{1}{\pi}\int_0^1 \frac{v(t)}{t-x}\,dt + \int_{-1}^{+1} H(x,t)v(t)dt = \Gamma(v(t))$$

where $v(t)$ denotes the pressure on the plate, $H(x,t)$ is an integrable kernel and Γ is a well behaved operator.

Equation (5) was solved (see Butler et al) by first writing

$$v(x) = \frac{u(x)}{\sqrt{1 - x^2}}$$

and expanding $u(x)$ as a Chebychev series. The coefficients of this series were then evaluated using collocation.

This work has aided the understanding of transducer calibration.

7.3.3 Coccidia in Chickens

Coccidia in chickens is an intestinal disease caused by protozoan parasites of the genus Eimera which costs the poultry industry some $250 - $300m per annum worldwide in drug treatment. Unilever Research had the idea of including coccidia in the bird's feed with a view to stimulating their immune response. However, although the Company developed the technology to achieve this (Coccidia would normally die outside a water environment) it proved to be expensive and it was not clear what quantity of coccidia need he included. At this point it was suggested that a mathematical model of the ecology of the parasite might well yield the precise amount of the vaccination required.

An interdisciplinary team was set up consisting of three immunologist and three mathematicians including Professor J.D. Murray as a consultant. The life cycle of the coccidia is shown in Figure 6.

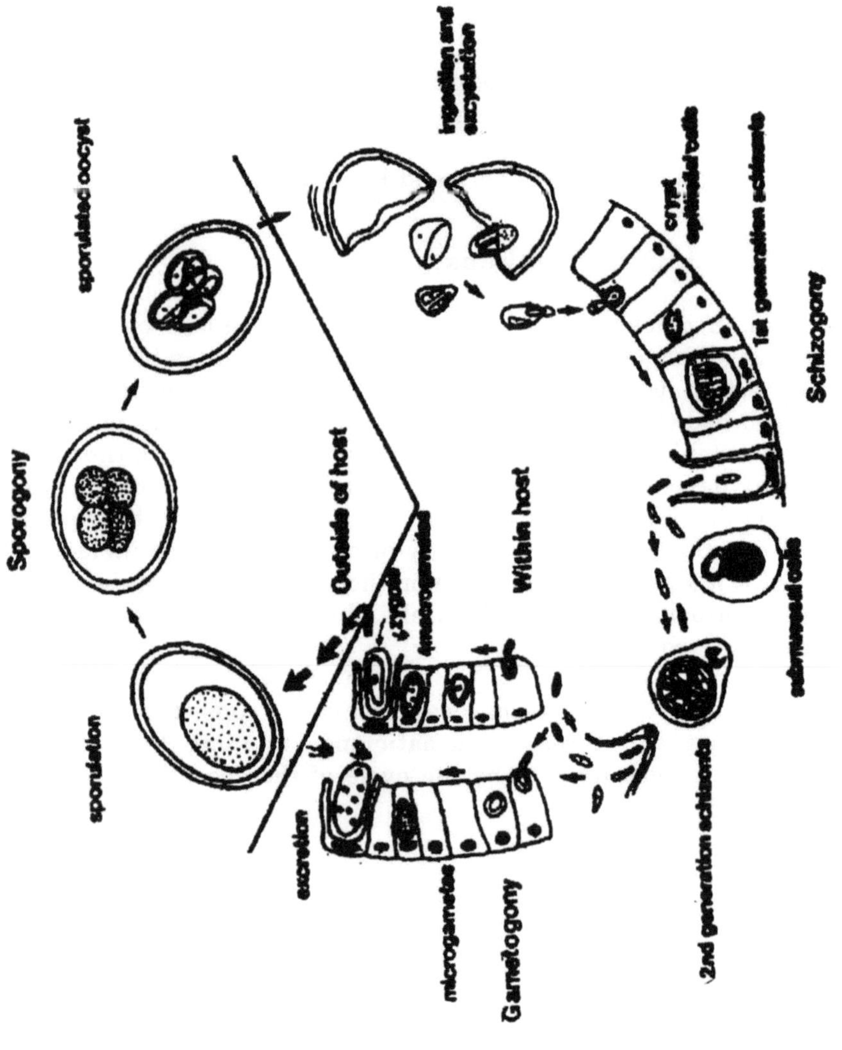
Sporogony
sporulated oocyst
sporulation
Outside of host
Within host
ingestion and excystation
crypt epithelial cells
1st generation schizonts
Schizogony
submucosal cells
2nd generation schizonts
excretion
microgametes
Gametogony

Figure 6.

After a substantial number of interdisciplinary meetings and discussions over many months the following model emerged:

$$\dot{x}_0 = a_1 x_4(t - t_1) - d_0 x_0 - b_0\{w(t)/w_0\}x_0$$

$$\dot{x}_1 = F(t) + b_0\{w(t)/w_0\}x_0 - b_1 x_1$$

$$\dot{x}_2 = x_1 - b_2 x_2$$

$$\dot{x}_3 = x_2(t - t_2)/\{1 + I(t - t_4)\}^p - b_3 x_3$$

$$\dot{x}_4 = x_3(t - t_3) - b_4 x_4$$

where

$$I(t) = \int_0^t K(s)x_3(s)ds$$

and

$$K(t) = K_0\{w(t)/(w(t) + w_0)\}^2$$

where

a_1 is a re-scaled sporolation rate constant

b_i is the re-scaled death rate constant

K_0 is the immune constant

and $w(t)$ is the weight of the bird as a function of time.

This system of delay integro-differential equations was solved using an implicit scheme with good stability properties and such that, by solving the equations in a specific order, it admitted a solution explicitly.

Dec screen management software (SMG) was used to provide a user-friendly menu-driven front-end, while menu-driven graphical output enabled the user to quickly view his results. This has meant that the biologist with little or no mathematical background has been able to use the program, and in particular to experiment and check that the results obtained from the model agree with their general knowledge and experience of the field, information which may not have been specifically articulated at an earlier stage.

This software has aided the discovery of the optimal amount of coccidia to include in the bird's feedstuff and the best way to administer it - by a steadily accumulating dose. As a result it seems extremely likely that the chickens we shall be eating in the next year or two will be vaccinated in this way.

7.3.4 Rotor stacking of gas turbine components

High frequency vibrations are detrimental to the lifespan of gas turbine engines such as those used, for example, in jet aircraft. One source of vibrations is a mass imbalance of the engine components. These components (rotors, stators and spacers) may be thought of as right circular cylinders which are joined together in a prescribed order by means of tie rods through a circular pattern of bolt holes in the top and bottom faces of each component.

Owing to these imperfections in manufacture. the rotor faces may be slightly skewed with respect to the walls; also, the pattern of bolt holes may not be exactly centred on the rotor face. As a result of the skews and eccentricities, the centre of mass of individual rotors in the assembled turbine will not lie exactly on the bearing axis of the turbine, by which I mean the straight line joining the centres of the top and bottom faces of the assembled stack of rotors (Figure 7). Their mass imbalance has led to vibration during the operation of the turbine, and it is important to minimise this.

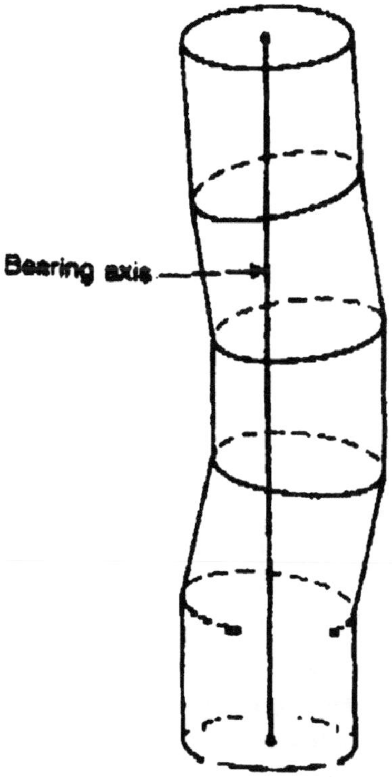

Figure 7.

These mass imbalances can, however, be minimised by choosing which hole on the top of one rotor is to abut which hole on the bottom of the succeeding rotor. There are sixteen holes and, coincidentally, sixteen rotors. Enumeration would require 15^{16} attempts (since the bottom rotor may be regarded as fixed).

By considering a sequence of axes defined on the centre of the bottom face it is possible, through elementary three dimensional analytical geometry, to express the distance of any point from the bearing axis of the assembly stack. The following minimisation problem is arrived at (see McKee and Reed (1987)):

$$\min_{\substack{\theta_j \\ j=1,\dots,m}} \max_{i \in \{1,\dots,N\}} \{\Delta_i\}$$

where $\Delta_n = c_n^T c_n - c_n^T \omega / \omega^T \omega$

where $c_n = \left[\prod_{i=1}^{N-1} B_i \right] d_n + c_{n-1}$

where $B_i = A_i^T R_{i+1}^T$

with $c_1 = d_1 = (\varepsilon_1, \delta_1, l)^T$.

Here

$$R = \begin{bmatrix} \cos\theta & -\sin\theta & 0 \\ \sin\theta & \cos\theta & 0 \\ 0 & 0 & 1 \end{bmatrix}$$

and

$$A = \begin{bmatrix} \cos\phi & 0 & \sin\phi \\ -\sin\phi\sin\psi & \cos\psi & \cos\phi\sin\psi \\ -\sin\phi\cos\psi & -\sin\psi & \cos\phi\cos\psi \end{bmatrix}$$

Here c_n is the coordinates of the top centre of the nth rotor with respect to the coordinate system of the bottom-most rotor and d_n is the coordinates of the top centre of the nth rotor with respect to the coordinate system of the bottom face of that rotor. (Note ε_n, δ_n denotes the skew and eccentricity of the nth rotor, while l is simply its length which is assumed to be constant.) Finally Δ_n denotes the distance of the centre of the top face of the nth rotor from the bearing axis of the complete assembly.

The problem has now been reformulated as one of minimisation of a nonlinear function of several variables (the angles θ_i) taking integer values. As such this problem is, of course, NP complete (Garey and Johnson (1979)). However, an algorithm based on the branch-and-bound technique has been derived and, when use is made of the fact that the skews and eccentricities are small so that second order terms may he neglected, has been demonstrated to be feasible (McKee and Reed (1987)). When, after a standard amount of flying time, certain rotors require replacing this provides an automated means of minimising any mass imbalance in the new assembly.

7.4 Concluding Remarks

This talk has brought out a number of points. Firstly, the mathematician collaborating with **private** industry (and note that all problems discussed came from **private** industry), particularly if the industry is multi-faceted and not inclined to employ mathematicians, needs to be a generalist who is an able and enthusiastic problem solver capable of using whatever mathematics is necessary. Equally, he must be able to work in an interdisciplinary team, and this often involves a painful learning process. Nevertheless, it is always easier for the mathematician to pick up the rudiments of immunology than for the immunologist to learn about functional delay differential equations.

Collaborating with industry can be very rewarding. This paper has, I hope, brought out a diversity of mathematical applications. Collaborating with industry can also help to feed back new ideas for the less adventurous, but possibly more cerebal, applied mathematicians to study in greater depth over a longer time scale and so advance the field of applied mathematics.

How should an academic collaborate with industry? I believe this is best done as a consultant which might involve spending short periods at the company. It is vital to know your client's business, and to understand the priorities of his problems. Working with your client builds confidence and eventually can lead to large contracts, but it does take time.

Collaborating with industry is rather like wooing a beautiful woman: it requires a great deal of perseverance and not a little flattery. Certainly, they are impressed if you are always chasing them.

7.5 References

[1] Burgess, N., Dixon, J., Jones, S. and Thoma, M. (1986). A reaction-diffusion study. UCINA report series (Ed. S. McKee), report 86/2.

[2] Butler, D., Cuminato, J.A. and McKee, S. Sound scattering from an underwater finite flexible plate. Math. Eng. Indust. (to appear).

[3] Garey, M.S. and Johnson, D.S. (1979). Computers and Intractability. San Francisco: Freeman

[4] McKee, S. and Reed, M. (1987). An algorithm for the alignment of gas turbine components in aircraft . IMA J. Maths Manag. 1, 133-144.

[5] McKee, S. and Stokes. A. (1983). Product integration methods for the nonlinear Basset equation . SIAM J . Numer. Anal . 20, 143-160.

[6] Parry, S., Barret, M., Jones, S., McKee, S. and Murray, J.D. The development of the immune response to coccidia in chickens (to be submitted).

[7] Swailes, D. and McKee, S. (1988). Graph matching and gel electrophoresis. Proceedings of the third European Conference for Mathematics in Industry (to appear) (eds. J. Manley, S. McKee and D.H. Owens). University of Strathclyde, Scotland.

8. Industrial Mathematics: General Remarks and Some Case Studies

H. Neunzert, Universitat KaisersLautern

Industrial mathematics has many faces; but its essential feature is the cooperation of partners - from industry and from universities - with quite different interest (business versus academic carreer), normally working on different time scales. They measure success in a different way (selling rate against citing index), they have different hierachies of values and are very often distrusting each other. Industry doubts that mathematicians are willing and/or able to produce something real practical and useful (and the mathematicians should not be too much surprised about this attitude, they very often doubt themselves) - mathematicians are afraid to loose their competence (their ideal of scientific truth, to say it more idealistically), to sell their souls.

The only way to convince both sides of the usefulness of such a messalliance is to show examples, which satisfy both: Examples with practical success, where "real" (i.e. publishable) mathematics is involved - even better: where real mathematics has to be invented. As an ideal, at least from a mathematical point of view, one might consider the following remark, which Euler made in a letter from 1736 [1] : "Therefore you may see, most Honourable Sir, that this solution according to its character has almost no connection to mathematics, and I do not understand why such a solution should be expected more from a mathematician than from any other human being. The solution is based only on reason, and principles of mathematics are not necessary in order to find it". The solution he mentions was that of the Konigsberg seven bridges problem, the origin of topology.

Our mathematical dreams are certainly not identical with the dreams of those, who pose us the problems - this was even true for Euler, about whom the Prussian king Frederic the Great was complaining in a letter to Voltaire: "The lifting tackle was built according to mathematical calculations and as yet cannot lift a single drop of water up to 50 feet from its container. Vanity of vanities! Vanity of Mathematics!". The origin of many disharmonics between the partners is a different meaning of the word "solution". Solution for mathematicians means: Existence and uniqueness - a qualitative explanation of a typical behaviour - a numerical algorithm; solution for people in praxis today means very often: Software; they want an instrument for checking an old, designing a new system, they have no time to study carefully "the interior" of such an instrument, it must be reliable and easy to handle. I believe that this idea of what a solution should be is held by a strong majority - in spite of the fact that "the industry" does not exist, as does not "the mathematician". Just to name two extremes: Aeronautical industry and the computational fluid dynamist are partners, who have very little problems with each other. But there are many sometimes quite small companies producing specialities with a highly developed very sophisticated technology, who never thought about mathematics - in spite of the fact that mathematics is hidden everywhere in their production process. And there is on the other hand the pure mathematician, who cannot see any application of his knowledge - in spite of the fact that he would be quite happy if such an application would exist. There are many of these small companies and many pure mathematicians - and both would gain something in a partnership, if they would dare to try.

That's somehow the policy my "Arbeitsgruppe Technomathematik" (Laboratory for Technomathematics) at the University of Kaiserlautern follows. Educated as a pure mathematician and starting the academic carreer with existence theorems for equations of mathematical physics, 15 years ago I became interested in the question what mathematics means outside the university, what our students were doing when leaving the university after graduation. Travelling through German industry of all kind and asking for open problems which might be solvable by mathematical methods (and not selling my special competence), I established a cooperation with until now more than 30 companies, most of them relatively small but also including for example Siemens, AUDI and Marcel Dassault. Cooperation comprises education and research: Companies are posing problems for student seminars and for master thesis - or we make research contracts with them. Contract means that industry pays the salary for the researcher - money is in this case an honest measure for their interest. Taking money certainly creates the danger to become dependent, to do only simple programming or classified work. Being aware of this danger makes it easy to avoid it (for example one should insist in making the mathematical content of the research publishable but should accept to keep all information about the industrial partner secret).

In order to illustrate the kind of problems, which satisfied both partners in being mathematically nice and practically successful, I shall now present two examples taken from our work during 1988.

8.1 Quality Control for Artificial Fabrics (see [2])

Artificial fabrics are sometimes produced by air spinning processes. Several hundreds of plastic fibres are drawn by wind, at the end moving turbulently and falling on a rolling ribbon (see Figure 8); there they stick together and a texture is produced, which has a huge variety of applications.

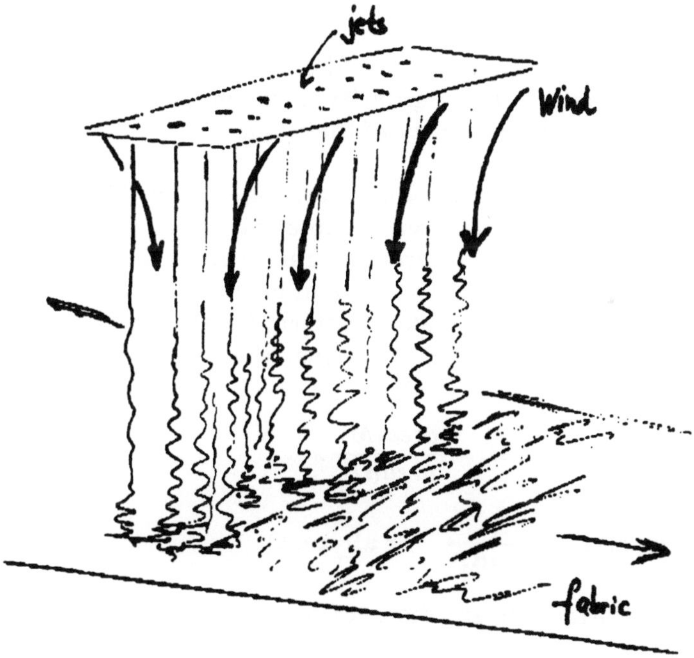

Figure 8.

There are many aerodynamical problems involved - some of them extremely difficult. For example: Make a mathematical model for the interaction of fibres and flow, which explains the dependence of the fibre thickness on the air speed and which describes the fluttering of the fibres as elements of the whole ensemble. We are working in this direction but do not have enough results to report on it. But there is a much simpler problem connected: The tissue produced always shows some defects, irregularities which are called clouds (spots of several centimeters diameters which are darker or brighter than the average in through-shining light) or stripes (an anisotrpy in the fibre structure, which appear when several fibres stick together); these defects become more severe, when the plant, i.e. the wind channel becomes polluted during the production. What is needed is a quantitative measure for the irregularity, which can be calculated online and may therefore be used for quality and production control. One may assume that there is a laserbeam used, which gives the absorption through the texture on 4 parallel lines along the tissue. What is delivered are 4 functions showing the absorption along these lines, functions similar to stochastic time functions. But we are not interested in the statistical properties of these functions, but only in the regularity of the material; therefore we need a concept for the uniformity of the tissue.

Consider a piece Ω of the fabric as a pattern, given by the absorption: If M is a subset of Ω, $\mu(M)$ will denote the power of the light passing through M. Since we are only interested in deviations from uniformity, we can normalize $\mu(\Omega) = 1$. μ can be considered as a normalized Borel measure on Ω. The ideal pattern, i.e. the uniformity of the fabric is given by $\mathring{\mu}$, where $\overset{\Omega}{\mu}(M)$ is just the proportion of the area of M compared to the area of Ω

$$\mathring{\mu}(M) = \frac{area(M)}{area(\Omega)} .$$

$\mathring{\mu}$ has the density $\mathring{f}(x) = \dfrac{1}{area(\Omega)}$ for all $x \in \Omega$.

Deviation from uniformity means distance between μ and $\mathring{\mu}$.

$$\text{Irregularity of } \mu = I(\mu) = \text{ distance } (\mu, \mathring{\mu})$$

Which distance? μ and $\mathring{\mu}$ are normalized measures - which concepts for a distance does the theory offer?

To make things simpler (and more realistic), assume that Ω is segmented into N pixels and μ is just given by the values $\mu_1, \dots, \mu_N$ in these pixels. We have $\mu_i \geq 0$ and $\sum_{i=1}^{N} \mu_i = 1$.

The corresponding $\mathring{\mu}$ is $\mathring{\mu}_i = 1/N$ for all $i = 1, \dots, N$.

Now we have the following possibilities;

(a) Relative Entropy, where

$$\text{distance } (\mu, \mathring{\mu}) := E(\mu/\mathring{\mu}) := \sum_{i=1}^{N} \mu_i \ln \frac{\mu_i}{\mathring{\mu}_i} = -\ln N + \sum_{i=1}^{N} \mu_i \ln \mu_i$$

This distance, suggested by statistical physics is in fact a measure for nonuniformity - but not the appropriate one for us: $E(\mu/\mathring{\mu})$ is not "influenced by order" and this means here: One "big hole" in the fabric counts as many "small holes".

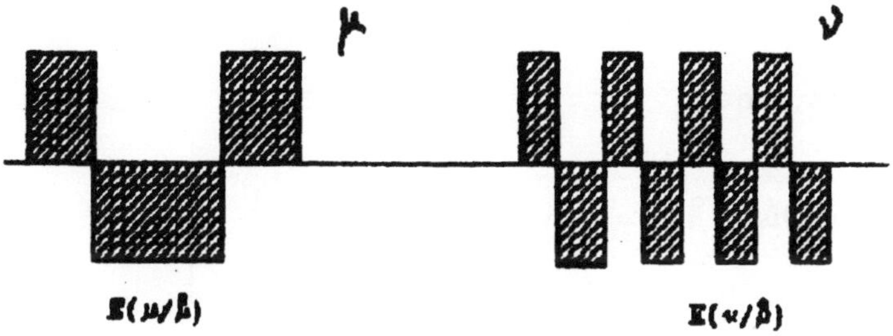

Figure 9.

This does not correspond to the companies idea of quality.

(b) Bounded Lipschitz Distance, where

$$\text{distance } (\mu, \mathring{\mu}) = D(\mu, \mathring{\mu}) := \max_{\underline{a} \in D} \left| \sum_{i=1}^{N} a_i \left(\mu_i - \frac{1}{N} \right) \right|$$

with $D := \left\{ \underline{a} \in \mathbb{R}^N \,/\, |a_{i+1} - a_i| \leq \frac{1}{N} \right\}$.

One realizes that the definition of this metric corresponds to a quite strange norm in $\mathbb{R}^N$

$$\|\mu\|_D := \max_{\underline{a} \in D} \left| \sum_{i=1}^{N} a_i \mu_i \right|$$

This distance filters somehow the holes (like a car driving over a road with holes on a certain speed), so that large holes are worse than many small ones. Mathematically speaking, D generates the weak topology in the space of probability measures - which means practically that small pattern shifts create small distance deviations.

D is a reasonable measure, but a little bit elaborate to be computed numerically. G. Rote ([3]) realized that the dual problem (in the sense of linear programming) is a curve smoothing problem allowing a quite simple and fast algorithm.

Simpler and in general as good is the

(c) Discrepancy, where

$$\text{distance } (\mu, \mathring{\mu}) = Dis(\mu, \mathring{\mu}) := \max_{1 \le \alpha \le \beta \le N} \left| \sum_{j=\alpha}^{\beta} \left(\mu_i - \frac{1}{N} \right) \right|$$

This concept originates from numbertheoretic consideration already in papers by Ilermann Weyl (1916) (see also [4]) and is related to the Kolmogorov-Smirnov distance in statistics. Again Rote [3] proposed a very simple algorithm:

Put

$$f_j := \mu_j - \frac{1}{N}, \quad j = 1, \ldots, N, \quad F_k := \sum_{i=1}^{k} f_i, \quad F_0 = 0.$$

Then

$$Dis(\mathring{\mu}, \mu) = \max_{0 \le i < j \le N} |F_i - F_j| = \max_i (F_i) - \min_i (F_i).$$

The algorithm works as follows:

$$\max := 0\,, \quad \min := 0\,, \quad F := 0$$

for $k = 1$ to N do $F := F + f_k$.

$$\text{If } (F > \max): \ \max := F$$
$$\text{If } (F < \min): \ \min := F$$

Then $Dis(\mathring{\mu}, \mu) = \max - \min$.

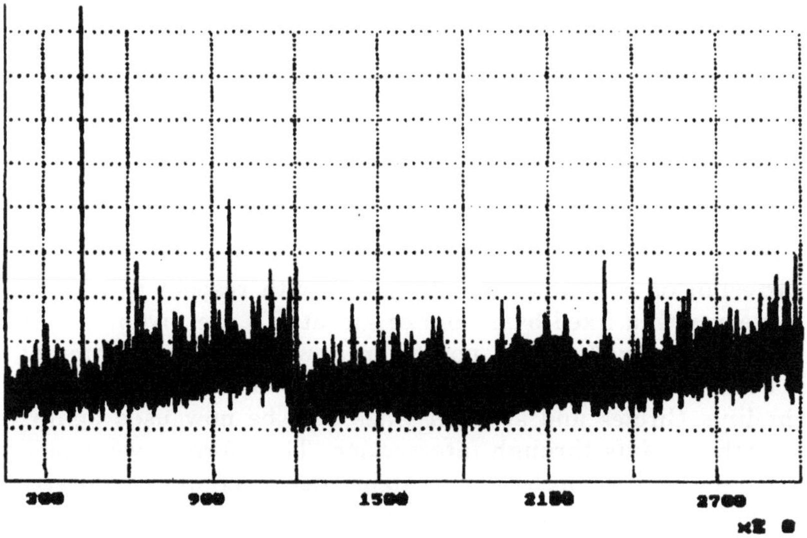

Figure 10.

Figure 10 shows discrepancy measurements along only one line of laser scanning; each value of $Dis(\mu, \mathring{\mu})$ is calculated for Ω being a ca. 1 m long piece of this line (which means that Ω is one dimensional, which we had implicitly assumed already in our definitions; these definitions can be easily extended to 2 or more dimensions - but not the algorithms!). Each abscissa in the diagram corresponds to such a piece of the line, each ordinate to the corresponding discrepancy. The discrepancy is in average increasing until the channel is cleaned ($x \approx 1200$) .

Discrepancy proves to be in general a useful measure of irregularity, even with only one scanning track. This is especially true, if clouds are the main source of defects. If stripes are more important, one has to define a measure for anisotropy - here different methods, maybe something like an 2d-wavelet analysis, should be tried. This work is in progress.

8.2 Computer Aided Design with Worst Case Tolerances (see [5] and [6])

The problem arose in car industry: Construction drawings of gears are normally given with tolerances for the sizes of the construction elements. A construction drawing consists of finitely many points in the plane together with straight lines and circles determined by these points and some additional information like the radii of the circles. These points are constructed recursively in using simple elementary constructions.

"Chains" of points $P_0, P_1, \ldots, P_N$ can be recursively constructed, if the (cartesian or polar) coordinates of P_j with respect to a coordinate system with a fixed basis and origin at P_{j-1} are given; $P_0 = 0$ is fixed "general origin". Each point can now be used as center of a circle with given radii, two points can be used to determine a straight line. Circles and straight lines may be now used again to define further points through intersection. Therefore, if we consider only construction by circle and straight edge, a point is recursively defined by previously given points through the following data:

(a) the coordinates with respect to a coordinate system with another point as origin,

(b) as intersection of straight lines and/or circles defined by pairs of points and by point and radius respectively.

The problem of tolerances is now easily defined: The data (coordinates, radii) are subject to given tolerances. Therefore the positions of the points are not uniquely defined. One may think of probability distribution instead of points - but the recursion leads quickly to hopelessly complicated formulae for these distributions, so that only simulation can be considered, if the number of points is more than say 10. Moreover, in the case of the gears, one is mainly interested in worst case considerations: Which positions of the point are in accordance with data and tolerances?

Instead of points P we consider tolerance sets $Tol(P)$, which are also depending on the tolerance sets for the previously constructed points, the data and their tolerances and certainly the construction method (i.e. (a) or (b)).

As example we consider only constructions without circle. In (a), the point P_1 is given with respect to fixed origin 0 with $Tol(0) = \{0\}$ by his coordinates. We write here $Tol_0(P_1)$ - certainly a rectangle for cartesian coordinates but a set like in Figure 11 for polar coordinates.

If a point Q is given with respect to a coordinate system, whose origin P is again in a tolerance set, then we get

$$Tol(Q) = Tol(P) \oplus Tol_0(Q)$$

where $A \oplus B$ denotes the Minkowvski sum $\{a + b / a \in \Lambda, \;\; b \in B\}$.

For chains we get

$$Tol(P_N) = Tol_0(P_1) \oplus \dots \oplus Tol_0(P_{N-1})$$

Figure 11 shows such a tolerance set with $Tol(P)$ as the interior of a circle and $Tol_0(Q)$ as the tolerance domain for polar coordinates.

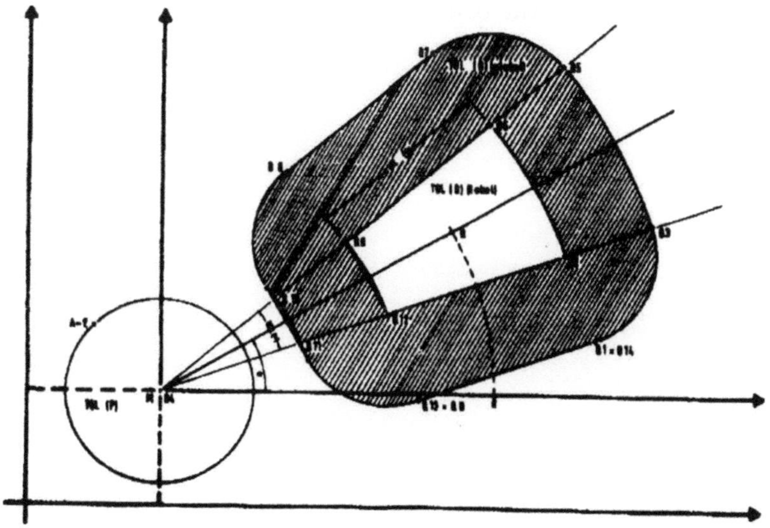

Figure 11.

One may realize that the computational effort will soon increase enormously. The essential idea now is to enlarge the tolerance domains a little bit by taking the convex hull as new tolerance set. This changes $Tol_0(P)$ for P given by polar coordinates. The enlargement is not very significant and since the Minkovski sum of convex sets remains convex, everything else discussed above has not to be changed.

Let's now consider straight lines g, defined by the tolerance set for two points P_1, P_2. One get again sets, which may be denoted by

$$Tol(g) := \{x/x \text{ lies on a straight line, which intersects } Tol(P_1) \text{ and } Tol(P_2)\}.$$

$Tol(g)$ is not convex and is not convexified. The construction (b) defines a point P as intersection of two straight lines g and h, and we may try to define

$$Tol(P) := \text{convex hull of } Tol(g) \cap Tol(h).$$

Figure 12 shows what may happen: $Tol(g) \cap Tol(h)$ is a convex or nonconvex polyeder, maybe unbounded, even disconnected. In the first two cases $Tol(P)$ is really defined as above; if $Tol(g) \cap Tol(h)$ becomes unbounded, the construction is not acceptable, a CAD programme will inform the user about the rejection.

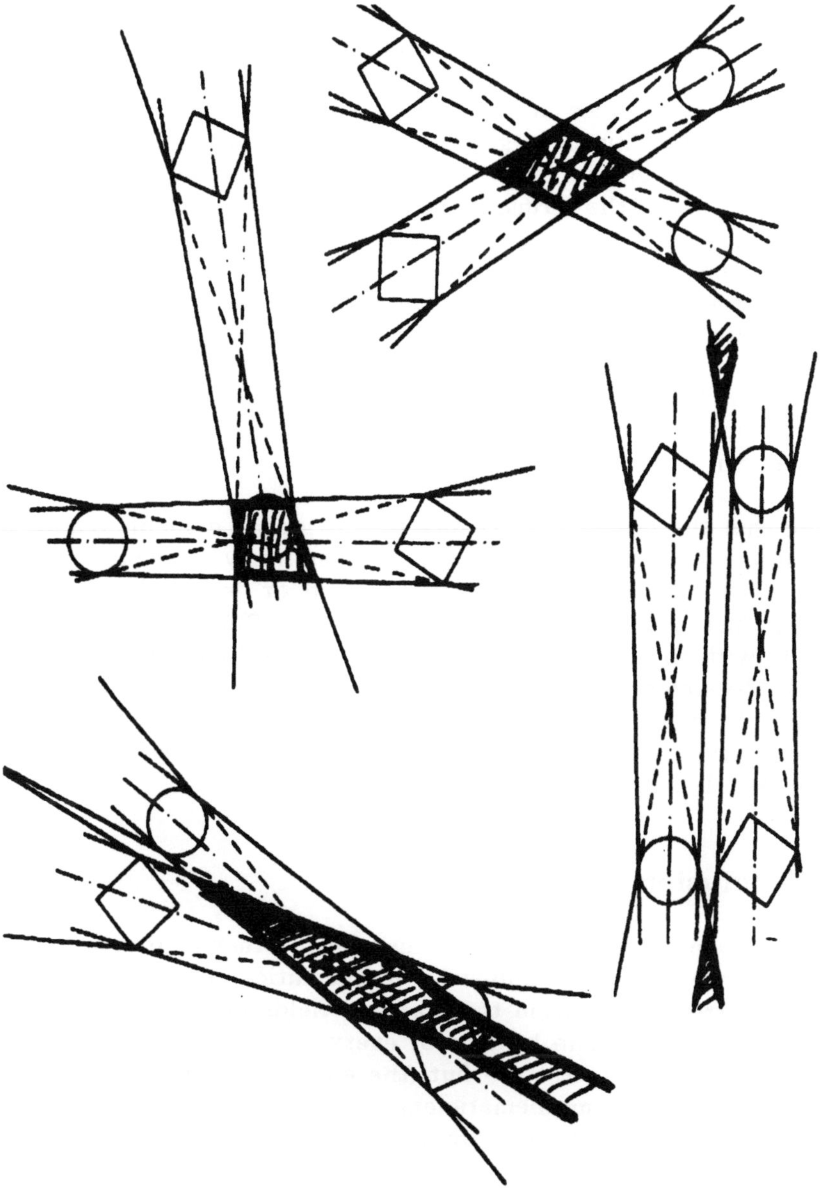

Figure 12.

In any case, the computational effort lies in determining the convex hulls and the Minkovski sums. But since the set is polyedric (or, if we use circles, "quasipolyedric", with boundaries consisting of pieces of straight lines or circular arcs), it is enough to find the extremal points. This is in general a linear programming task; the extremal points of the Minkovski sum are given as sums of the extremal points of the terms, if the interiors of the normal cones at those points have nonempty intersection. There is some convex geometry involved - not deep, but nice - and one finds that there are very quick simple linear programming algorithms sufficient to do the computations. The state of art now is, that we have an algorithm and a simple CAD programme to determine the tolerance domains for 2d construction drawings, if circle-circle are excluded. To complete the task means some programming effort (even new rejection criteria will become necessary).

I think that the practical relevance of this idea is evident. Where is the mathematical fun? Now, there have been many attempts to solve the problem of toleranced CAD, all more or less failing, if the drawing has a high complexity, since the effort increases very strongly with the numbers of points (for our method, it's essentially linear). Moreover, one can do a little bit axiomatic geometry - with convex sets instead of points etc. The situation is similar to what is called interval arithmetic - a theory much further developed but maybe even less rich and exciting in structure.

8.3 Conclusion

The two examples given above are chosen in order to illustrate the flavour of what we call industrial mathematics. The mathematics there is borrowed not from the classical fields of applications like pde, statistics etc.; it can be done by every mathematician, whatever his strict field of interest is - but (the engineers may apologize) it can be done only by mathematicians. Therefore it's worthwhile for the industrialist to discuss with mathematicians - they may be a source of innovation. And its worthwhile for the mathematician to consider the problems posed by industry - they give fun, prestige, even some money for his group. It's quite rare for scientistis that fun and profit goes together.

8.4 References

[1] Cited from H. Sachs / M. Stiebitz: "250 Jahre Graphentheorie" in NTM-Schriftenv. Gesch. Naturwiss. Technik, Med. Leibzig 24 (1987) 2, 2. pp. 57-62 (translation by N. Barton).

[2] H. Neunzert, R. Wetton: Pattern recognition using measure space metrics, preprint no. 28, Arbeitsgruppe Technomathematik, Universitat Kaiserslautern, 1987.

[3] G. Rote, personal communication, 1988.

[4] L. Kuipers, H. Niederreiter: Uniform Distribution of Sequences, 1974, Wiley, New York.

[5] U. Hinze: Erstellung eines mathematischen Modells zur Berechnung von Toleranzbereichen, Master Thesis, Kaiserslautern, 1988.

[6] U. Hinze, H. Neunzert, D. Socolescu: CAD for construction drawings with worst case tolerances, to appear.

9. Study Groups with Industry

J. R. Ockendon

Secretary MSGI

Mathematical Institute Oxford

9.1 Introduction

In any "mathematics - in - industry" scheme, a crucial feature which permits the deployment of mathematical talent is the identification of novel and exciting mathematical ideas in practical problems. Any scheme will fail to use mathematics to the full unless this identification takes place.

One way in which mathematicians can "gravitate" towards problems of interest is to encourage about 50 theoreticians to confront say 6-10 problems for a week long period and this is the format of the Study Groups which have been held in UK annually since 1968. Further details of these meetings, and of the problems described below, can be obtained from the author.

The problems discussed briefly in this talk are chosen not only for their mathematical interest but also to illustrate the impact that mathematics can have on industrial problems. A brief review of this aspect will be given at the end of the talk. An early example of the unexpected appearance of mathematical novelty was

9.2 Pantograph Dynamics

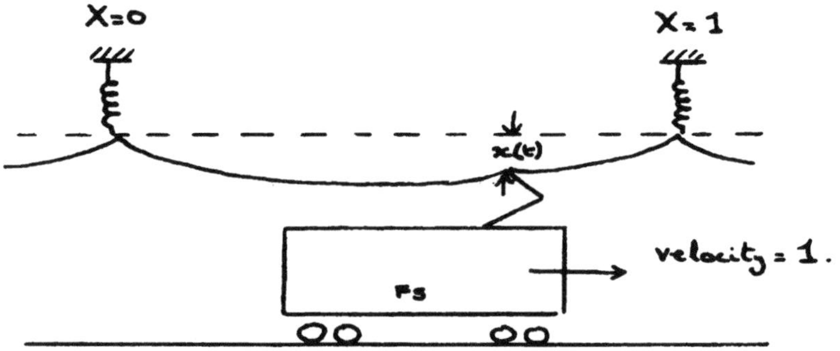

Figure 13.

The transverse wave speed in the overhead wire is $c > 1$ and the characteristics in the (X, t) plane are

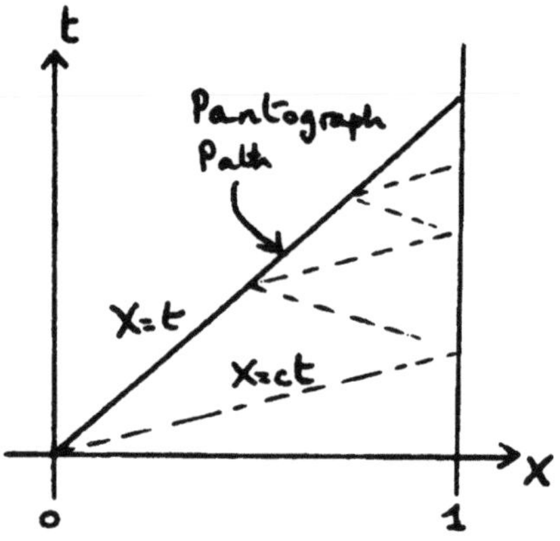

Figure 14.

Assuming a linear differential relationship between the force on the pantograph and the displacement $x(t)$, the "time-delay" between the emission of waves by the pantograph and the reception of waves reflected from the overhead supports means that

$$\dot{\underset{\sim}{x}}(t) = A\,\underset{\sim}{x}(\lambda t) + B\,\underset{\sim}{x}(t)\,, \quad \underset{\sim}{x}(0) = \underset{\sim}{x}_0$$

where $\underset{\sim}{x}$ is a vector whose first component is x. A ,B are constant matrices and $\lambda = (c-1)/(c+1)$. This functional differential equation has the possibility of oscillatory solutions, even in the scalar case. (T.Kato , J.B. McLeod; Bull Am. Math. Soc. 77 1971 p. 891-937)[2]

9.3 Thermal oscillations in pipes.

At a very recent "Study Group" in Madrid, the problem arose of gravity driven oscillations in vertical pipes with heat sources and sinks.

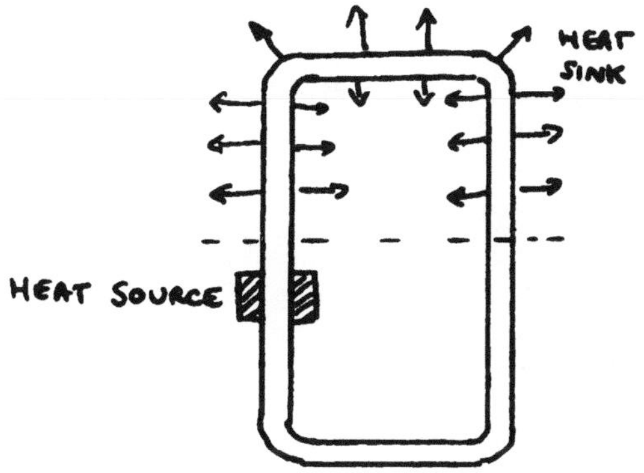

Figure 15.

The hydrodynamics in the pipe is simply modelled as a balance between buoyancy forces, pressure gradients and turbulent wall resistance. Thus one-dimensional energy and momentum equations can be written down. When the stability of the steady state is analyzed, the spatial periodicity of the flow in the pipe loop results in a "global" generalization of a Sturm-Lionville problem

$$u'' + \left(a + b\int_0^1 k(\xi)\, u(\xi)\, d\xi\right) u = 0$$

2 Other novel functional differential equations have arisen at the RPI workshops in connection with the design of perfectly focusing lenses.

with $u(0) = u(1) = 0$, where a and b are nonlinear functions of an eigen parameter and K a prescribed kernel.

More difficult modelling problems occur in

9.4 Synthetic Fibre Manufacture

It is observed that when, say, cellulose or acrylic fibres are spun from a into a hot bath,

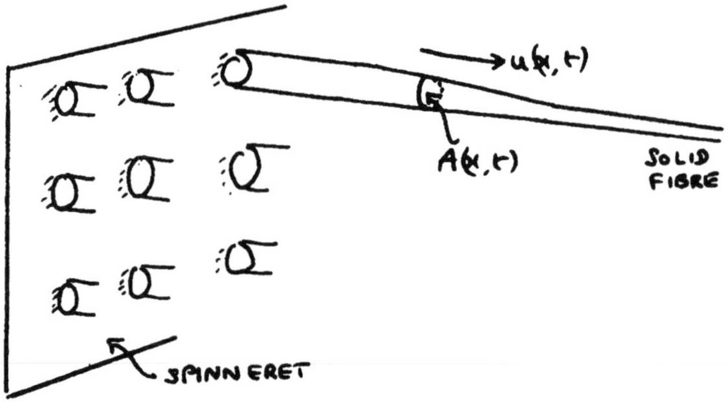

Figure 16.

the fibre velocity u is piecewise linear in space:

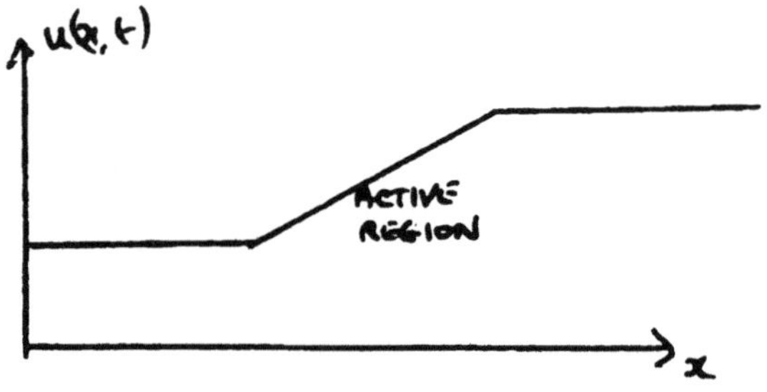

Figure 17.

Hence we may propose a model in which the area of the fibre cross section is A and, by

mass conservation $\quad \dfrac{\partial A}{\partial t} + \dfrac{\partial}{\partial x}(Au) = 0\,;$

energy conservation $\quad A\dfrac{\partial \theta}{\partial t} + Au\dfrac{\partial \theta}{\partial x} = h(\theta_0 - \theta)$

where $\theta =$ fibre temperature (h, θ_0 are constants); and momentum conservation gives the tension

$$T = T(t)\,,$$

where $T = \dfrac{\lambda}{\theta + \text{const}}$ in an "active" region where the strain λ is changing, and $T < \dfrac{\lambda}{\theta + \text{const}}$ elsewhere (where $\lambda =$ const).

Thus, if we consider incompressible fibres,

$$A\lambda = 1$$

and an explicit solution[3] is available in the active region which yields

$$\frac{dT}{dt} = R(T)$$

where R is a known rational function. This equation predicts instability if T is sufficiently small.

Numerical problems were posed by the study of

3 The corresponding problem for an incompressible liquid (e.g. glass) fibre is

$$\frac{\partial A}{\partial t} + \frac{\partial}{\partial x}(Au) = 0\,, \quad \frac{\partial}{\partial x}\left(3\mu A\frac{\partial u}{\partial x}\right) = 0$$

and can also be solved explicitly.

9.5 Gas Flow in Blast-Furnaces

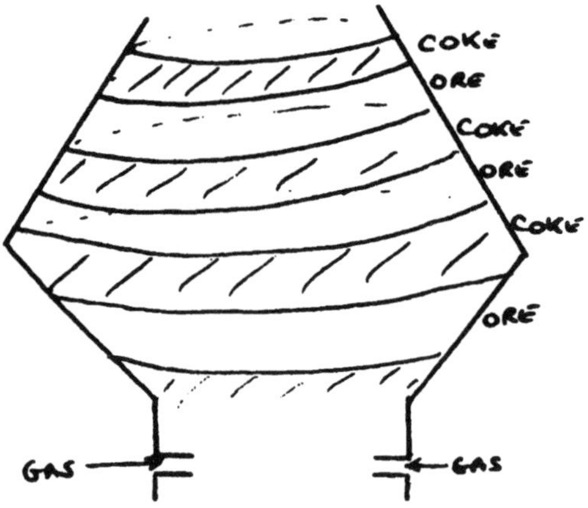

Figure 18.

Here the combustion gas flows at such high velocity that the Darcy law must be replaced by the Ergun equation

$$- \underset{\sim}{\nabla} p = (\underline{\underline{k}}_1 + \underline{\underline{k}}_2 | \underset{\sim}{V} |) \underset{\sim}{V} ,$$

where p is the gas pressure, $\underset{\sim}{V}$ is the gas velocity and $\underline{\underline{k}}_i$ are discontinuous function of position. We may now observe that if $\underline{\underline{k}}_i$ are scalars, we can invert to give

$$\underset{\sim}{V} = \frac{-2 \underset{\sim}{\nabla} p}{\left\{ k_1 + \sqrt{k_1^2 + 4k_2 | \underset{\sim}{\nabla} p |} \right\}}$$

and hence that a finite element discretization is possible which minimizes

$$J = \int \left\{ - \frac{6k_1}{k_2} | \underset{\sim}{\nabla} p | + \frac{1}{k_2^2} (k_1^2 + 4k_2 | \underset{\sim}{\nabla} p |)^{\frac{3}{2}} \right\} d\underset{\sim}{x}$$

An interesting historical twist is provided by the modelling of

9.6 Thermistors

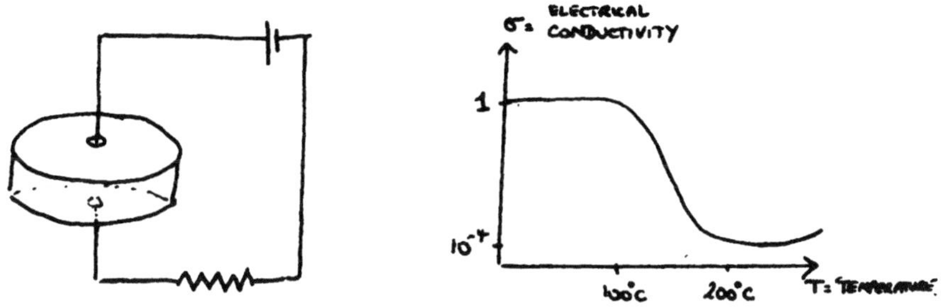

Figure 19.

Thermistors are electronic devices whose conductivity variation with temperature makes then act as fuses as a result of Joule heating. When the electric potential is φ , the steady state is modelled by

$$\underset{\sim}{\nabla}(\sigma(T)\ \underset{\sim}{\nabla}\varphi) = 0$$

$$\underset{\sim}{\nabla}(K(T)\ \underset{\sim}{\nabla}T) + \sigma(T)\,|\underset{\sim}{\nabla}\varphi|^{2} = 0\,, \qquad K = \text{ thermal conductivity.}$$

Typical boundary condition for a two-dimensional device are

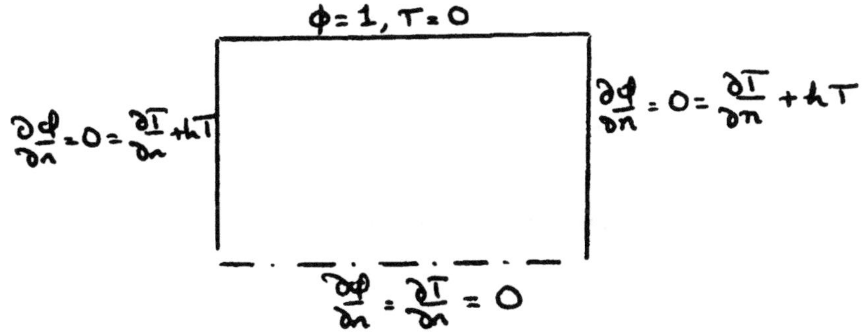

Figure 20.

As pointed out by Cimatti, Diesselhorst (Ann. Phys. 1 (1900) p. 312-325 has shown that when $h = 0$, $T = T(\varphi)$ i.e. the isotherms coincide with the equipotentials, and if we define ψ such that.

$$\underset{\sim}{\nabla}\psi = \sigma(T)\, \underset{\sim}{\nabla}\varphi ,$$

then $\Delta\psi = 0$ with

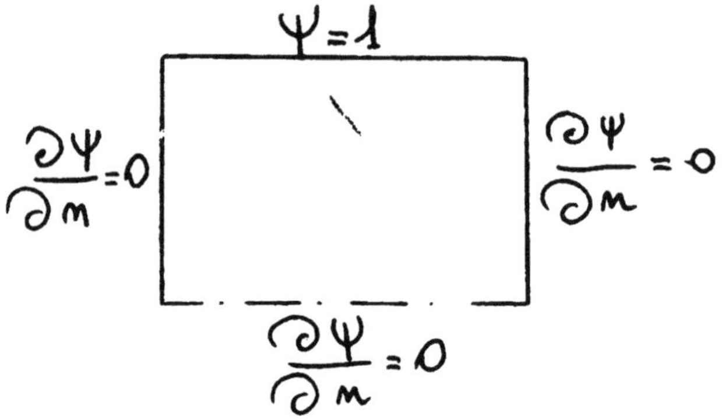

Figure 21.

Also, since T is given by

$$\int_0^T \frac{K(T')\, dT'}{\sigma(T')} = -\frac{1}{2}(\varphi^2 - 1) ,$$

there is the possibility of nonexistence of a steady state if K/σ is too small. For realistic σ the morphology is that of a very thin hot region (> 200°C) near the centre line of the device.

One of the largest class of mathematical models in industrial problems is that of

9.7 Stefan Problems (cf. the presentation of M. Primicerio)

These problems certainly arise in the study of welding (laser, resistance,...,), solidification (casting , ingots, alloys,...), food processing, oxidation, corrosion, semiconductor manufacture and device modelling, thermal switches, porous media flow, evaporation and commutator manufacture (all of which have been considered at Study Groups).
Their mathematical interest is apparent from the simplest "Stefan" model (proposed by Lamé and Clapeyron, Ann. de Chimic et Physique 47, 1832 p. 250-256).

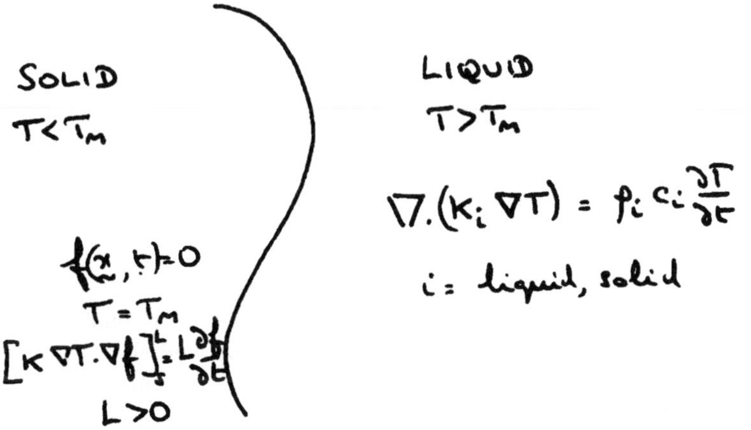

Figure 22.

and the "weak" formulation is

$$\frac{\partial E}{\partial t} = \nabla(K\nabla T) \quad \text{where } E = \rho\int_{T_n}^{T} c\, dt + LH(T - T_n)$$

and H is the Heaviside function. With suitable initial and fixed boundary conditions, local existence and uniqueness has been proved by Harzawa Tokoku Math. J. 33 (1981) p. 297-335 and also by Meirnarov Math. Sb. 112 (1980) p. 170-192. However, most problems in industry are supercooled ($T_{liquid} < T_n$) in which case the Stefan model is ill-posed. It needs to be regularized and the most likely mechanism is surface energy (i.e. the Gibbs-Thomson term) in which case $T_n = -\sigma\kappa$ where is a positive constant and κ is the near curvature of $f = 0$ with a suitable sign.

In this case the only available analysis is for the "Hele-Shaw" specialization in which a "one-phase" model is posed for $T_{liquid} = p$, with $c_{liquid} = 0$:

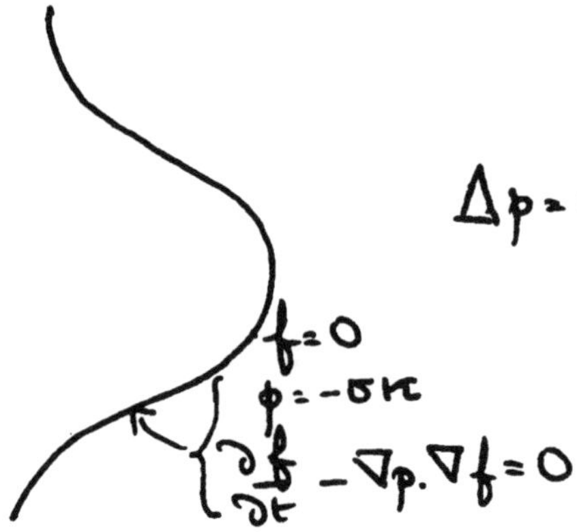

Figure 23.

the fixed boundary conditions are such that $p < 0$ at least in part of the liquid region. For this problem, it can be proved that when $\sigma = 0$ singularities develop in the form of cusps in the boundary:

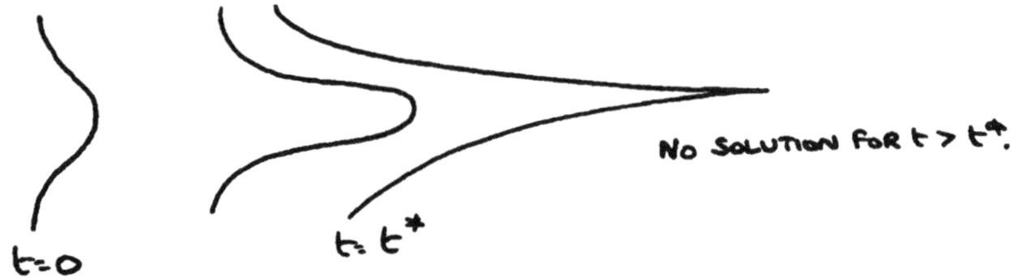

Figure 24.

However it is possible to conjecture that for $1 >> \sigma > 0$, and $t = t^* + 0(\sigma^{1/3})$, the morphology is "fractal":

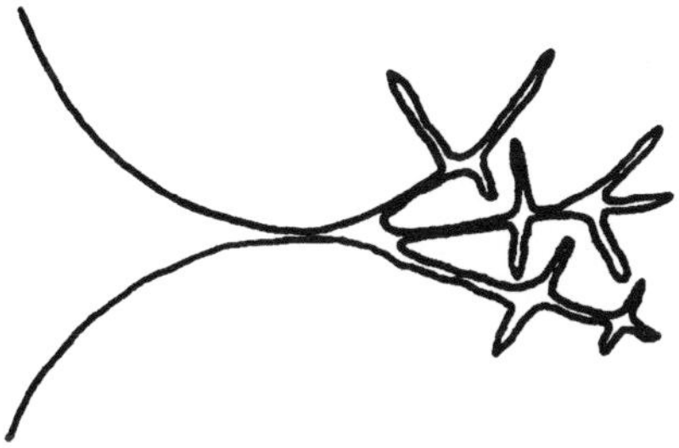

Figure 25.

9.8 Conclusion

In conclusion, we have presented a variety of Study Group problems which have both physical and mathematical interest. From the industrial viewpoint, the analysis of problems 3 and 4 has proved to be especially beneficial, whereas problems 1, 5 and 6 have stimulated much mathematical research.

The future seems to be one of an ever increasing demand for industrially-oriented mathematics, at least as long as the computee revolution persists. This means a continuing need for an environment in which mathematicians can exercise their imagination in teams and experience with Study Groups suggest this is best done a) when financial considerations are not unduly obtrusive and b) when the industrial collaboration persists for several days at least.

10. Some Free Boundary Problems in Industrial Mathematics

M. Primicerio

Universita' degli Studi

Istituto Matematico Ulisse Dini

Firenze - Italy

The aim of this talk is to present some examples of free boundary problems which have been studied by our group, with reference to technological applications.

I will recall that free boundary problems are problems in which (part of) the boundary of the domain in which the problem is studied is among the unknowns.

The colleagues who were involved in Firenze in the researches I will describe are A. Fasano, M. Ughi, F. Gori, R. Ricci and E. Comparini.

I hope that the few examples I will describe - ranging from standard models to an essentially new problem - will give a general idea of our work in the area of the technological applications of mathematics.

10.1 Freezing of liquids in satured porous media

10.1.1 Practical problems

Among many examples we can recall:

(i) freezing of soils underlying paved roads and buildings foundations: the problem of frost heave;

(ii) thawing of permafrost due to man-made structures (e.g. the trans-Alaska pipeline);

(iii) artificial freezing to consolidate soils during tunneling, underpinning, archaeological excavations;

(iv) freezing of foodstuffs;

(v) cryosurgery.

10.1.2 The simplest approach (Stefan-type models)

In this model, liquid, solid and the porous matrix are supposed at rest and at each time t an unknown surface Γ of equation $\Phi(\underline{x},t) = 0$ is supposed to separate the region where the liquid in the porous medium is frozen from the unfrozen domain.

If θ denotes the temperature ($\theta = 0$ being the freezing point) and K and c denote the thermal conductivity and heat capacity respectively, and if we use suffixes U and F for unfrozen and frozen respectively, we have the following situation

$C_F \dfrac{\partial \theta_F}{\partial t} = \operatorname{div}(K_F \nabla \theta_F)$	Γ	$C_U \dfrac{\partial \theta_U}{\partial t} = \operatorname{div}(K_U \nabla \theta_U)$
FROZEN		UNFROZEN

Figure 26.

On Γ we will have

$$\theta_U = \theta_F = 0$$

$$\varepsilon \lambda \rho \frac{\partial \Phi}{\partial t} = - K_F \nabla \theta_F \cdot \nabla \Phi + K_L \nabla \theta_L \cdot \nabla \Phi$$

where λ is the latent heat, ε is the porosity and ρ is the density of the liquid (which is supposed to remain unchanged upon freezing, to be consistent with the assumption of neglecting movement).

We have tested this model by means of the apparatus shown on Figure 27.

The vessel was filled with a sample of a given sand saturated with a given liquid. The two thermostats were set at the freezing temperature of the liquid $+0.1°C$ for one day. Then the bottom was suddenly brought to a temperature $-\Theta$ below the freezing temperature and the movement of the freezing front was recorded by means of the thermocouples.

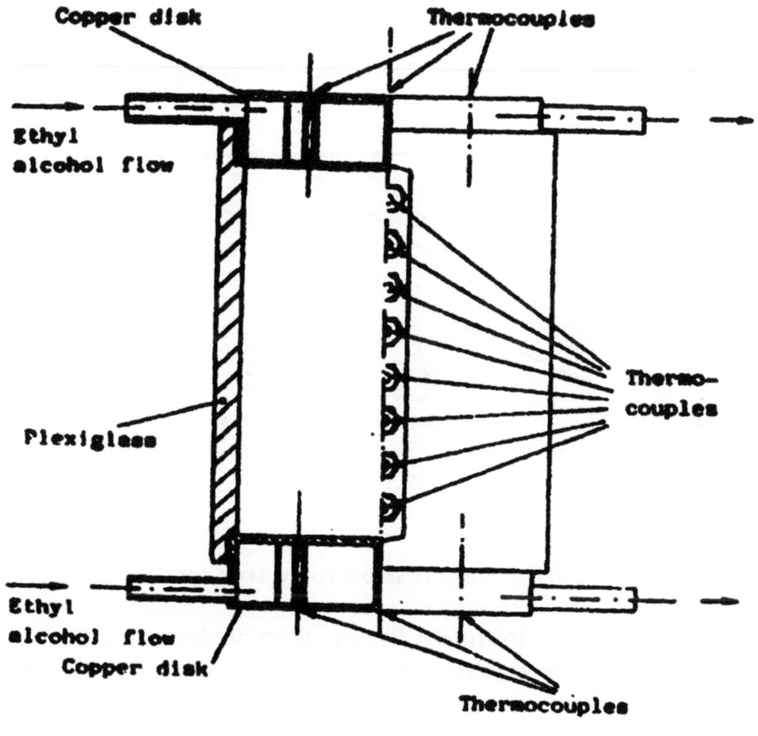

Figure 27.

The experimental situation simulated a one-dimensional, one-phase problem for which it is known that the position s of the front is given by

$$s = A\sqrt{t} \tag{1}$$

where

$$A = 2\gamma \frac{\sqrt{K_F}}{\sqrt{C_F}}$$

and γ is the solution of the trascendental equation

$$\sqrt{\pi}\ \gamma\, e^{\gamma^2}\ \mathrm{erf}\,\gamma = C_F \frac{\Theta}{\lambda}.$$

In Figure 28 it can be seen that the experimental points fit the expression (1) rather well, when the sand was rather coarse and the liquid used was water.

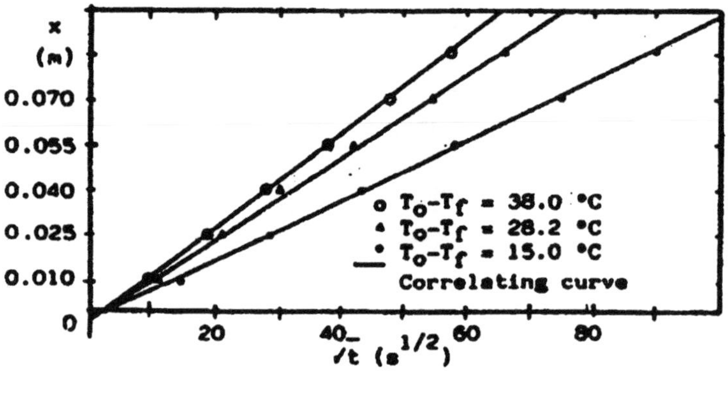

Figure 28. Freezing front penetration for water saturated sand

For finer sands and for liquids different from water, significant discrepancies were observed: the scheme fails when the pore diameter decreases and/or when non pure liquids are used. In particular the liquid content can change relevantly during the process (Figure 29).

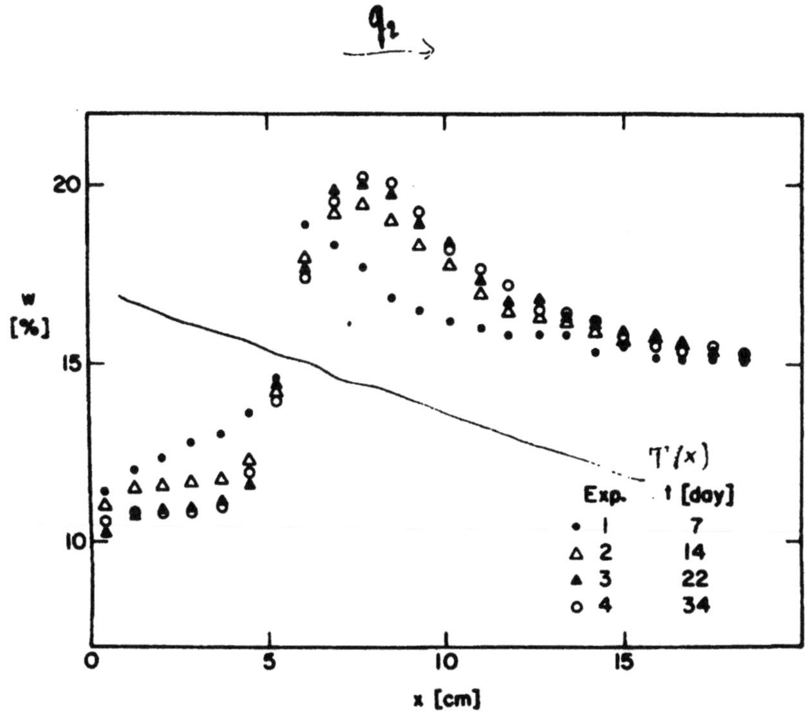

Figure 29.

10.1.3 Freezing of foodstuffs.

I will just recall that the mathematical definition of a weak solution to the Stefan problem sketched above is based on the introduction of a function

$$E(\underline{x}, t) = \int_{\theta_m}^{\theta(\underline{x}, t)} c(\eta)d\eta + \varepsilon g\lambda \mathrm{H}(\theta - \theta_m),$$

where θ_m is the melting temperature and

$$\mathrm{H}(z) = \begin{cases} 0, & z < 0, \\ 1, & z > 0. \end{cases}$$

Of course, E turns out to have the meaning of a thermal energy and the problem can be written as a single equation in the whole domain

$$(2) \qquad \frac{\partial E}{\partial t} = \text{div } (K \nabla \theta)$$

Since E and K (if $K_F \neq K_U$) have discontinuities at $\theta = \theta_m$ (see Figure 30 and Figure 31), equation (2) has to

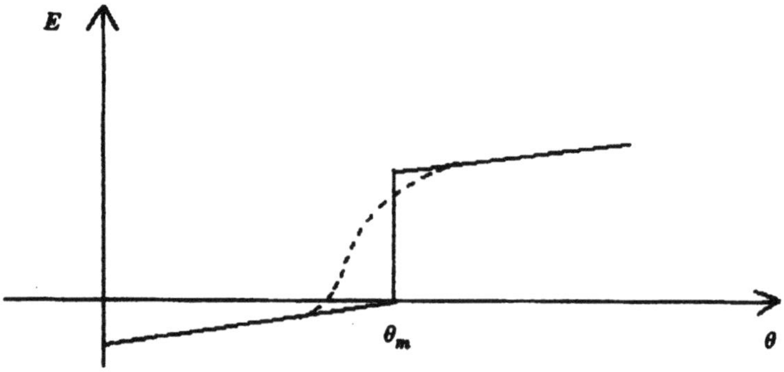

Figure 30.

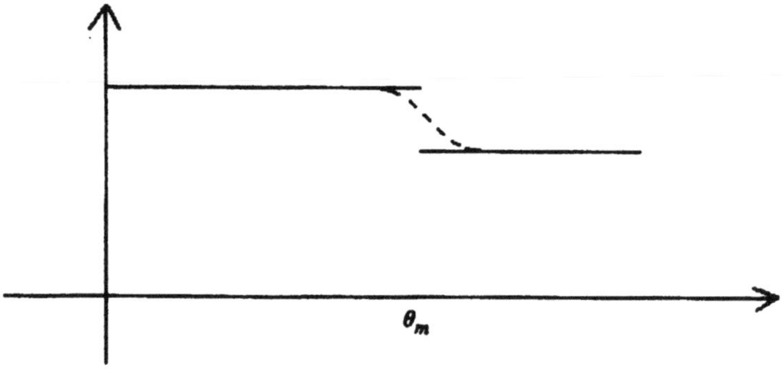

Figure 31.

be read in a suitable "weak" sense I will not define here for sake of brevity.

Very roughly speaking, it can be said that E and K are approximated smoothly over a interval whose width is let to tend to zero.

Dealing with foodstuffs and measuring K and E (or $dE/d\theta$ as in the figure below), one find the following profiles, which refer to a material called "Tylose", simulating rather well the thermal behaviour of some foodstuffs. (Figure 32).

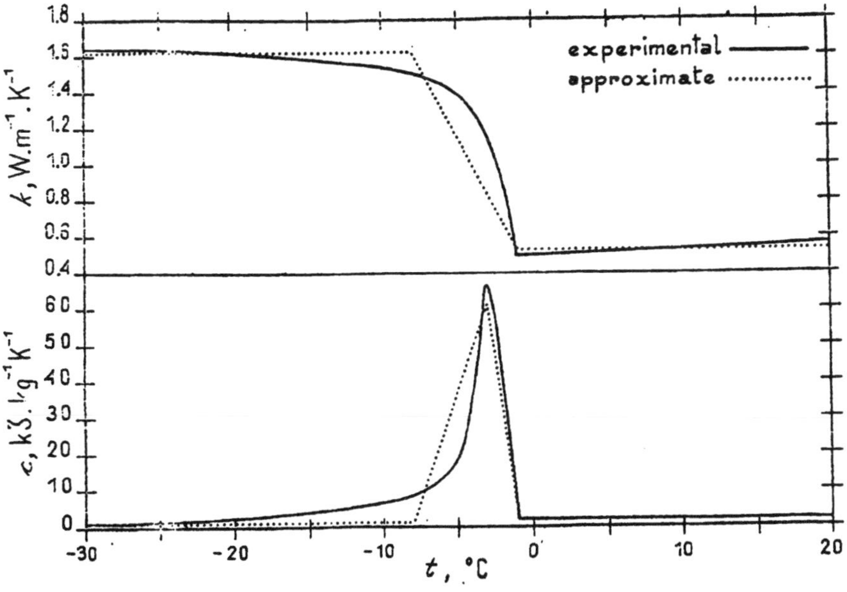

Figure 32.

Since it is hard to measure the exact shape of the peak around the freezing point, whereas the total area under it is easily found by calorimetric experiences, we used simple shapes - triangular, rectangular, etc. - with the same area and we provided a method of computing the change of phase of foodstuffs which is simple, accurate and based on rigorous mathematical analysis which will be of course omitted here.

The figures (Figure 33 and Figure 34) below show the comparision of our results (using a triangular approximation) with the experimental data corresponding to the solid curve of Figure 32.

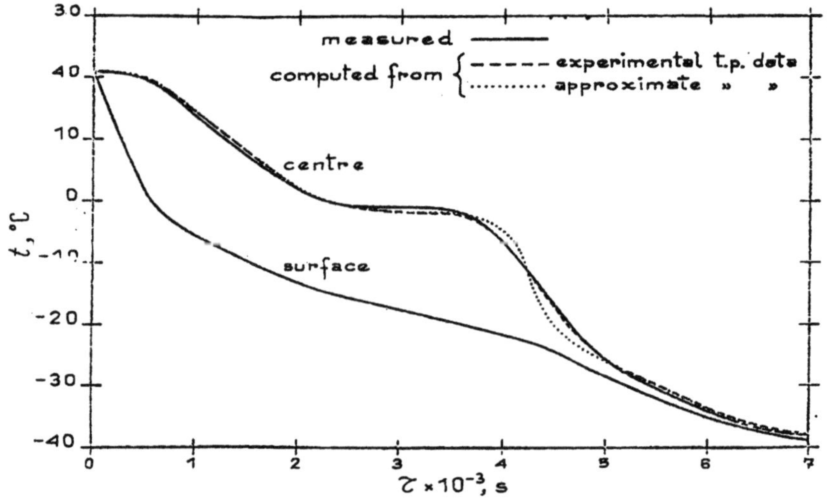

Figure 33.

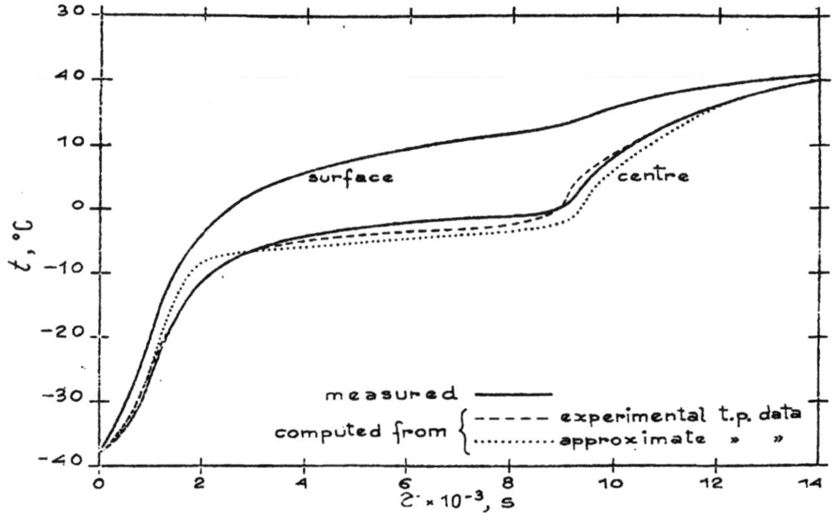

Figure 34.

Table 1 display some results obtained with real foodstuffs.

Table 1

t(°C)	Hamburger			Mashed potatoes		
	Surface	Centre		Surface	Centre	
$\tau \times 10^{-3}$(s)	Measured	Measured	Computed	Measured	Measured	Computed
0	19·3	19·3	19·3	20·1	20·1	20·1
0·24	10·1	18·8	19·3	10·2	19·8	20·0
0·72	−2·8	16·3	16·7	−1·9	16·1	16·9
1·20	−7·9	11·0	11·0	−5·3	10·0	10·9
1·68	−12·2	5·4	5·3	−8·6	4·9	5·3
2·16	−15·2	0·8	0·7	−11·4	1·0	0·8
2·64	−17·5	−2·1	−1·0	−13·8	−0·5	−1·3
3·12	−19·7	−2·4	−1·2	−15·9	−1·6	−1·8
3·60	−21·6	−3·7	−1·8	−18·1	−3·9	−2·6
4·08	−22·7	−9·5	−9·7	−20·3	−8·4	−4·2
4·56	−25·8	−19·7	−20·8	−24·8	−20·0	−21·7
5·04	−29·5	−26·9	−26·0	−29·9	−26·9	−27·5
5·52	−33·2	−31·1	−30·1	−32·8	−31·6	−31·6
6·00	−35·2	−33·6	−33·3	−35·3	−34·2	−34·3
6·48	−37·1	−36·2	−35·5	−37·2	−36·8	−36·4

10.1.4 Coupled heat and mass transfer in freezing soils

Measurements made using nuclear magnetic resonance show that in freezing soils liquid water and ice coexist over a wide range of temperature. Thus a relevant variable of the process is $w(\underline{x},t) \in [0,1]$, representing the fraction of unfrozen water.

Whenever $w < 0$ the liquid can move and denoting by $\underline{v}$ the velocity and by p the pressure a Darcy's type law

$$\underline{v} = -D\nabla p \tag{3}$$

is assumed to hold, but the hydraulic conductivity D depends on w, since formed ice becomes part of the porous skeleton. A satisfactory empirical law is

$$D = D_0 w^7 \tag{4}$$

(Note that gravity has been neglected just for sake of simplicity).

In the zone where $0 < w < 1$, Clapeyron's law gives

(5) $$\frac{P}{\rho} - \frac{P_i}{\rho_i} = \frac{\lambda\theta}{T_0}$$

where P, ρ and P_i ρ_i are the water and ice pressure and densities respectively, T_0 is the absolute freezing temperature and $\theta = T - T_0$.

Capillarity is accounted for by Kelvin-Laplace law

(6) $$P_i - P = \sigma\psi(w)$$

where σ is the surface tension and ψ is some averaged curvature of the water-ice interface whose dependence on w is experimentally found to be as in Figure 35.

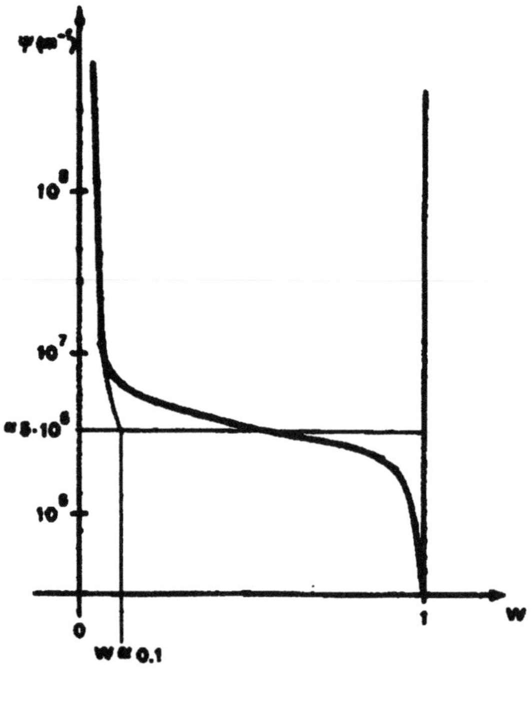

Figure 35. Freezing characteristic curve (thick line experimental)

In our model ψ is approximated by

$$\psi(w) = \begin{cases} \psi_0\left(\frac{w_0}{w}\right)^2, & 0 < w \leq w_0 \\ \psi_0, & w_0 < w < 1, \end{cases}$$

where ψ_0 and w_0 are measured constants (see the dotted line in Figure 35).

Just to save time, we will consider the model in the case $\rho = \rho_i$. (It can be shown that, the general case, the relevant quantities can be expressed as converging power series of $\delta = (\rho - \rho_i) / \rho_i$). Therefore (5) and (6) yields

$$\theta = - B\sigma\psi(w)\,, \qquad B = \frac{T_0}{\rho\lambda} \tag{7}$$

Consider a quasi-static one-dimensional problem in the layer $(0\,,L)$, with boundary conditions

$$\theta(0) = \theta_b > 0\,, \quad p(0) = p_0\,, \quad \theta(L) = \theta_a < 0\,, \quad p_i(L) = p_0\,.$$

The plane $x = s$ separating the liquid zone ($w = 1$) from the freezing fringe $x \in (s\,,L)$ is found to have temperature $\theta_0 = - B\sigma\psi_0$ and to be located at

$$s = L\frac{\theta_0 - \theta_a}{\theta_b - \theta_a}\,.$$

The water pressure is

$$P = P_0 - qx\,, \qquad 0 \leq x \leq s$$

where q is an unknown constant ($q = v/D_0$). Using Darcy's law one finds

$$q = - w^7\frac{dp}{dx}\,, \qquad s < x < L \tag{8}$$

whence (7) gives $dp / dz = - (q / w_0) \times (\theta^{7/2} / \theta_0^{7/2})$. Integrating this equation and imposing the boundary conditions the value of q is easily found

$$q = \frac{9}{2}\frac{\theta_b - \theta_A}{BL}\,W_0^7\left(\frac{\theta_b}{\theta_a}\right)^{\frac{7}{2}}$$

This expression for the water migration agrees with the empirical formulas proposed by experimentalists.

The model allows to find the ice pressure and hence to predict if and where, in a given soil under given condition, ice lenses will be found.

10.2 Flow problems of concentrated coal-water suspensions

10.2.1 The practical problem

To reduce transportation costs of coal, coal-water suspensions are prepared such that they are

(i) stable against sedimentation,

(ii) easily pumpable in long pipelines,

(iii) highly concentrated so to increase transportation efficiency and to make them directly burnable without dewatering.

This is achieved

(a) by finding the optional grain-size distribution (which is essentially bi-modal with peaks around 10 and 100 μ, approximately)

(b) by means of suitable additives (which are designed to prevent inbition of the particles to make it possible that all water is available as a lubricant).

It is found that the rhelogical properties of the suspensions tend to deteriorate under long pumping times and/or high velocity gradients; both situations are expected to occur in practical situations (over the long pipelines and in the burners).

10.2.2 Basic assumptions

In Figure 36 the experimental apparatus is shown in which flow tests have been performed at the laboratory of ENI Ricerche, Milano, in the framework of a joint research conducted by SNAM Progetti, ENI Ricerche and our group.

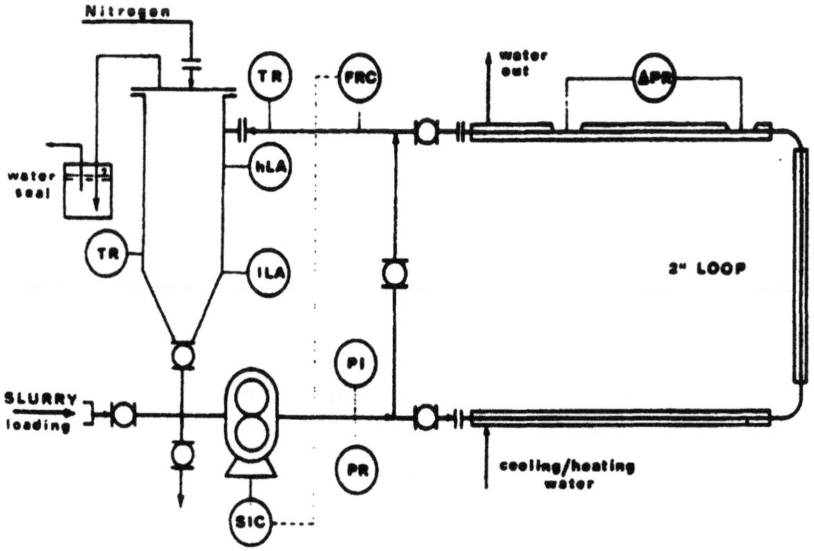

Figure 36. Test loop

In Figure 37 the ratio between the apparent viscosity η_A of the suspension and its initial value η_0 is recorded as a function of the pumping time on the test loop. The three curves correspond to three different pumping velocities corresponding to nominal shear rates of 20, 50 and 80 sec^{-1} respectively.

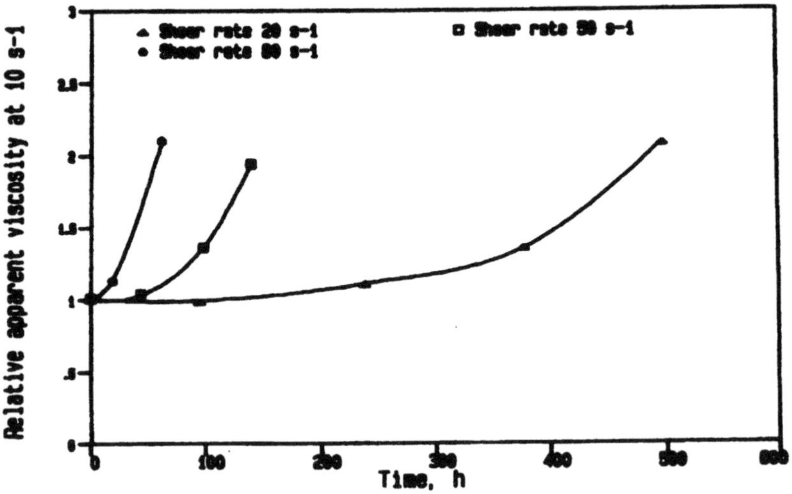

Figure 37. Effect of pumping time on slurry viscosity

Let us recall that if the relationship between the stress and the shear rate is expressed as

$$\tau = f(\dot{\sigma})$$

then, the apparent viscosity η_A is

$$\eta_A = \frac{f(\dot{\sigma}_0)}{\dot{\sigma}_0}$$

where $\dot{\sigma}_0$ is a reference shear rate (in the case under examination $\dot{\sigma}_0 = 10\ \text{sec}^{-1}$). Of course in the case of a newtonian fluid η_A is nothing but the viscosity of the fluid.

The assumption we took as a base for our model is that η_a is a function of the total energy E supplied to the suspension. Figure 38 supports this assumption showing that if η_A / η_0 is displayed as a function of E, the experimental points corresponding to the three tests above lie on the same curve.

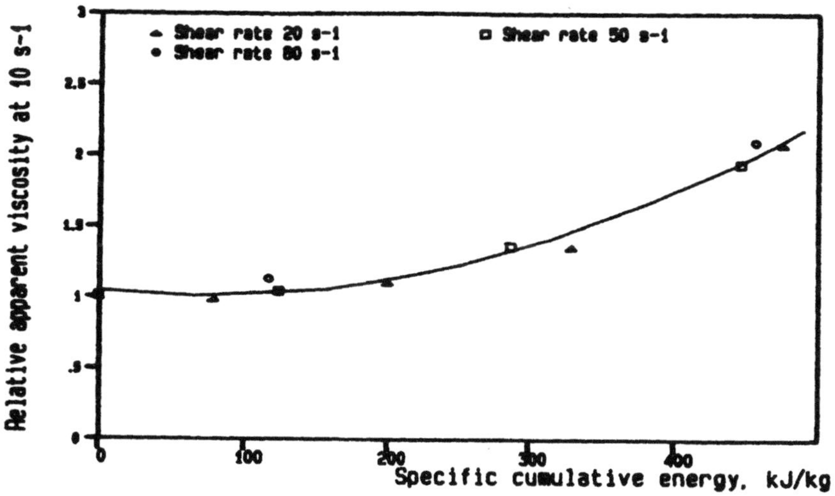

Figure 38. Effect of pumping energy on slurry viscosity

One more check cames from a comparison with an experiment in which the suspension is stirred in a modified rotational viscometer (Figure 39).

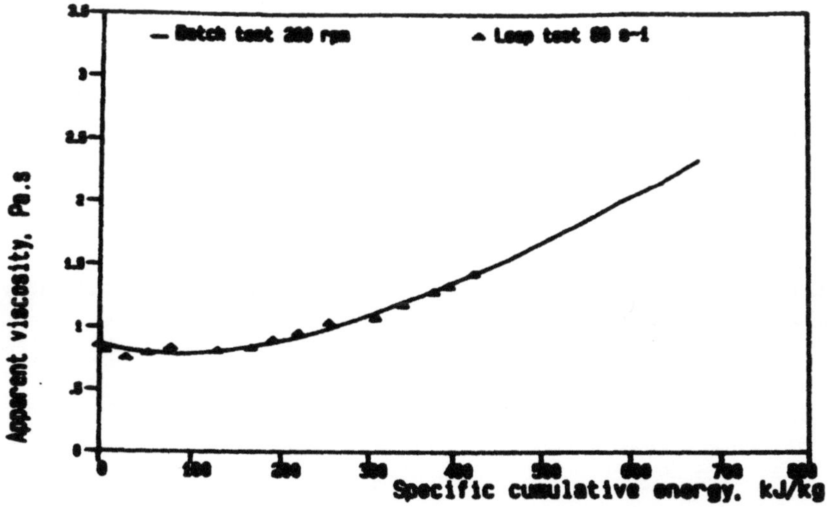

Figure 39. Degradetion profiles in batch and loop tests

10.2.3 A mathematical model

So far, no assumptions have been made on the from of $f(\dot{\sigma})$, i.e. on the rheological behaviour of the suspension. As a starting point, we consider a newtonian fluid, assuming that the rate of change of the viscosity is proportional to the energy dissipation rate.

Consider the simplest geometry in which the velocity is

$$\underline{v} = v(y\,,\,t)\,\underline{i};$$

we have the following pair of differential equations

(9) $$g\,\frac{\partial v}{\partial t} = \frac{\partial}{\partial y}\left(\eta\,\frac{\partial v}{\partial y}\right)$$

(10) $$\frac{\partial \eta}{\partial t} = \alpha\,\frac{\partial E}{\partial t} = \alpha\eta\left(\frac{\partial v}{\partial y}\right)^2$$

to be supplemented with the specification of the initial and boundary conditions.

A theorem of local well-psedness of the above mathematical problem has been proved. The result is global in time if time-scale analysis is performed showing that the term $\partial v\,/\,\partial t$ in equation (9) can be neglected.

It can be noted that (9), (10) admit a class of elementary solutions of the from

(11) $$\eta = \eta_0 \exp\left(\alpha\,\frac{V_0^2 t}{L^2}\right).$$

Figure 40 shows $\ln\frac{\eta_A}{\eta_0}$ as a function of time showing a surprisingly good agreement with (11).

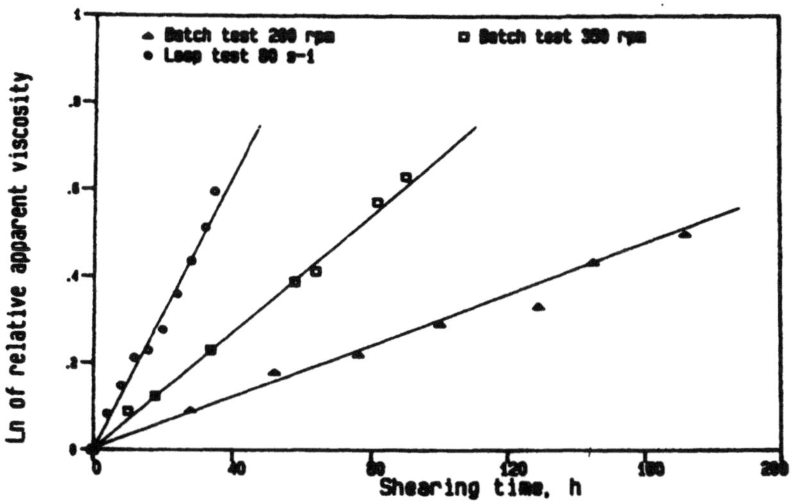

Figure 40. Semilogaritmic plot of relative viscosity vs. time

10.2.4 Modelling the suspension as a Bingham fluid

A more accurate description of the rheological behaviour of the coal-water suspension seems to be the Bingham model. Figure 41 displays on a $\tau - \dot{\sigma}$ diagram the behaviour of newtonian, pseudoplastic and Bingham fluids

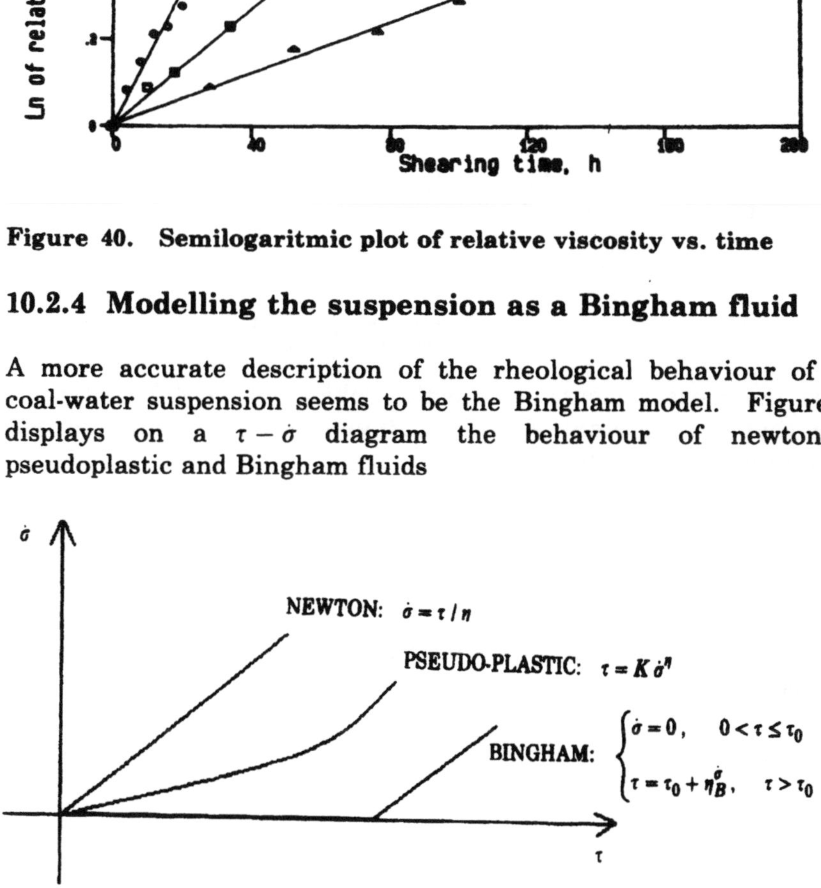

Figure 41.

Consider the stationary flow of a Bingham fluid in a tube ($\underline{v} = v(r)\underline{i}$, $v(R) = 0$) under a constant pressure gradient G.

It is easily found that, setting

$$r_0 = \frac{2\tau_0}{G} ,$$

we have the following velocity profile

$$v(r) = \frac{R}{\eta_B}\left(1 - \frac{r}{R}\right)\left[\frac{\tau_w}{2}\left(1 + \frac{r}{R}\right) - \tau_0\right], \qquad r_0 \le r \le R ,$$

$$v(r) = \frac{R}{2}\,\frac{1}{\eta_B \tau_w}\,(\tau_w - \tau_0)^2 , \qquad 0 \le r \le r_0$$

where τ_w is the measurable stress at the wall $r = R$. If U denotes the mean velocity, i.e. the total flow Q divided by the cross section πR^2 (all quantities are measurable) we find

(12)
$$\frac{R\tau_w}{4U} = \frac{\eta_B}{1 - \frac{4}{3}\frac{\tau_0}{\tau_w} + \frac{1}{3}\left(\frac{\tau_0}{\tau_w}\right)^4}$$

Since the measurements on samples taken out at different times show that $\eta_B =$ constant, equation (12) gives τ_0 as a function of time during the experiments.

The current research is aimed to ascertain if the increase of τ_0 is related to the energy dissipated over the whole cross section of the tube or if it is a local effect confined to the free boundary separating the "plug" $r \le r_0$ from the layer $r \in (r_0 ,R)$. A major point will be the study of the possible variation of r_0 as a function of t and its asymptotic behaviour.

10.3 References

[1] BONACINA C., COMINI G., FASANO A., PRIMICERIO M., Numerical solution of phase-change problems, INT. J. HEAT MASS TRANSFER, 16,1825-1832 (1973).

[2] BONACINA C., COMINI G., FASANO A., PRIMICERIO M., On the estimation of thermophysical properties in nonlinear heat-conduction problems, INT. J. HEAT MASS TRANSFER, 17, 861-867 (1974).

[3] FASANO A., PRIMICERIO M., Free boundary problems: theory and application, RES. MATH. NOTES PITMAN'S, Vol. 78-79, Pitman ,London,(1983).

[4] FASANO A., PRIMICERIO M., Freezing in porous media:a review of mathematical models, Proc. of German-Italian Synposium "Applicatins of Mathematics in Tecnology",V. Boffi- H. Neuzert eds., Teubner, 288-311 (1984).

[5] FASANO A., PRIMICERIO M., Heat and mass transfer in quasi-steady ground freezing processes, Proceedings ESMI, (M.Hazewinkel ed), Reidel -(1986).

[6] PRIMICERIO M., Dynamics of slurries Proceedings 2nd Symposium Math. in Industry, Teubner 1988.

11. Some Experience with Mathematics and Industry at INRIA

M. Robin

INRIA, B.P. 105, 78153

Le Chesnay Cedex, France

11.1 Abstract

A few examples of on-going programs involving INRIA and industry are described. Topics like combustion, fluid mechanics, vibrations control and identification of mechanical structures are concerned.

Contents

- Brief overview of INRIA activities
- Simulation of combustion problems
- Optimum design in fluids mechanics
- Control of a shocks absorber
- Identification and monitoring of mechanical structures
- Concluding remarks

11.2 Brief overview of INRIA activities

INRIA is the French national research institute in computer science and automatic control. Its role is to undertake research, to built experimental systems and simultaneously to maximise its output to industry.

INRIA activities cover a large part of computer science and applied math, control and systems sciences.

Over 50 teams are distributed among 4 laboratories (Nancy, Rennes, Rocquencourt and Sophia-Antipolis). The research topics can be described under the eight following programs:

1. Programming, symbolic computation and artificial intelligence

2. New computer architectures

3. Networks and distributed systems

4. Data bases

5. Automation, production automation, data and signal processing

6. Robotics, image and vision

7. Scientific computing, numerical software and computer - aided engineering

8. Man-machine communication

Obviously, a lot of mathematics is included in these topics and, a non exhaustive list is the following:

- nonlinear P.D.E.

- scientific computing

- control - optimization

- complexity

- stochastic processes
- logic and formal computation

In order to illustrate some projects carried out with industry, we choosed the following set of examples:

- Simulation of combustion problems
- Optimum design in fluids mechanics
- Control of shocks absorber
- Identification and monitoring of mechanical structures

These examples are by no means "finish products" but samples of on-going programs.

11.3 Simulation of combustion problems

Reference : B. Larrouturou et al. with SEP and CNES.

The following model is used to study the simulation of flame instabilities which requires performant numerical methods due to the stiffness of the problems.

The thermodiffusive model in an infinite duct:

$$T_t = \Delta T + \Omega(Y, T)$$

$$Y_T = \frac{1}{L_e} \Delta Y - \Omega(Y, T)$$

$T(x, y, t)$ (normalized) temperature

Y % reactant

$\Omega(Y, T)$ (normalized) reaction rate

$$\Omega(Y, T) = \frac{\beta^2}{2 \cdot L_e} Y \cdot \exp\left(\frac{-\beta(1-T)}{1-\gamma(1-T)} \right) = Y \cdot f(T)$$

(L_e Lewis number).

Initial conditions

$$T(x, y, 0) = T_0(x, y), \; Y(x, y, 0) = Y_0(x, y),$$

Boundary conditions

$$T(-\infty, y, t) = 0 \quad Y(-\infty, y, t) = 1$$
$$T(+\infty, y, t) = 1 \quad Y(+\infty, y, t) = 0 \quad \text{(reactant consumed)}$$

$$\frac{\partial T}{\partial n} = 0, \quad \frac{\partial Y}{\partial n} = 0 \quad \text{on } W \text{ (wall)}$$

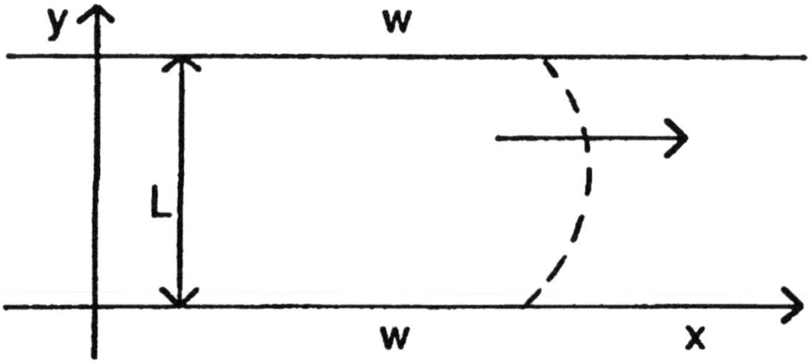

Figure 42.

Bounded domain for numerical simulation

$$D = [a, b] \times [0, L]$$

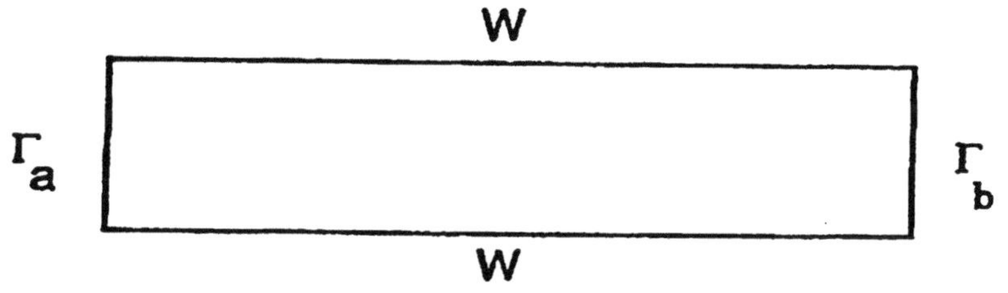

Figure 43.

D will be moving towards the fresh mixture (to follow the flame front), at a speed $V(t)$ close to average flame speed

$$V_0(t) = -\frac{1}{L}\int \Omega(Y\,,\,T)dx\,dy$$

(choose $V(t)$ such that $\int_D T =$ constant).

The model is now, in this moving reference frame:

$$\begin{aligned}
&T_t = \Delta T + \Omega(Y\,,\,T) + V(t)T_x \\
&Y_t = \frac{1}{L_e}\Delta Y - \Omega(Y\,,\,T) + V(t)Y_x \\
&\frac{\partial T}{\partial n} = \frac{\partial Y}{\partial n} = 0 \text{ on } W \\
&T = 0\,,\quad Y = 1 \text{ on } \Gamma_a \\
&\frac{\partial T}{\partial n} = \frac{\partial Y}{\partial n} = 1 \text{ on } \Gamma_b \text{ (hot boundary)} \\
&+ \text{ initial conditions.}
\end{aligned}$$

Main difficulty : when β is large this system becomes very stiff.

Classical analysis for $\beta \to +\infty$ (infinite activation energy)

One assumes the existence of a curve

$$x = \alpha(y\,,\,t)$$

$$\text{defining the flame front} \begin{cases} \text{separating the burnt region} & Y = 0 \\ \text{from the fresh mixture} & Y > 0 \end{cases}$$

One looks for solutions corresponding to periodic perturbations

$$\alpha(y, t) = Ae^{iky}e^{wt}, \quad A \ll 1.$$

The results are the following:

- **Stability** for $L_e \geq 1 - 2/\beta$
- **Instabilities** when $L_e < 1 - 2/\beta$ if the wave length $\lambda = 2\pi/k > \lambda_0(\beta, L_e)$.

Mesh adaptation

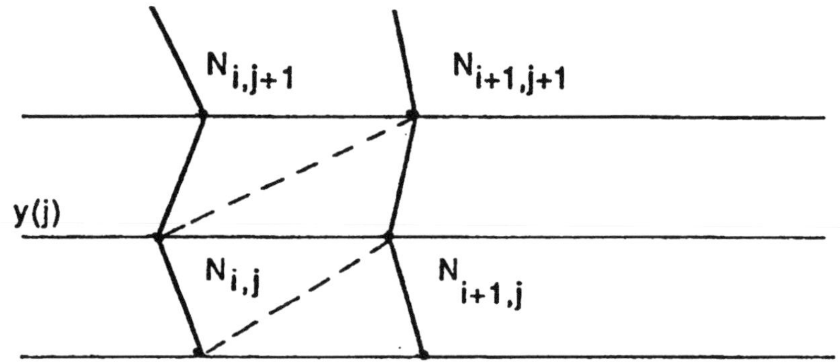

Figure 44.

In order to simulate numerically the previous model, an adaptive mesh strategy is used with a finite element method. At some point in time, the number of points in x is increased according to the following scheme:

for fixed j one changes the distribution of nodes

$$x^{\text{old}}(i, j) \rightarrow x^{\text{new}}(i, j).$$

One defines a mesh function w based on the truncation error of one dimensional equations, and x^{new} is determined by the criterium

$$\int_{x^{\text{new}}(i\,,j_0)}^{x^{\text{new}}(i+1\,,j_0)} w\,dx = \text{ constant } \forall i.$$

An analysis of this scheme is available. (cfr. [2], [3]).

Numerical experiments

1/. Instabilities - (uniform flow)

For $L_e < 1 - 2/\beta$ and a planar front as initial data, one can observe numerically that, after oscillations, stabilization occurs with a curved front (see Figure 45)

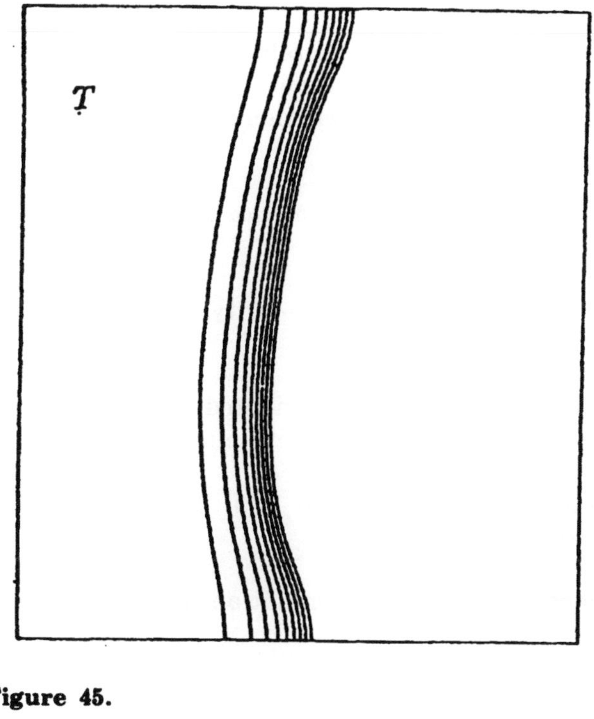

Figure 45.

2/. Stable curved flames

Curved flame propagation in non uniform gas flow. For this case, the model is the following:

$$T_t = \Delta T - K\cos\left(\frac{\pi y}{2L}\right)T_x + \Omega(Y, T)$$

$$Y_t = \frac{1}{L_e}\Delta Y - K\cos\left(\frac{\pi y}{2L}\right)Y_x + \Omega(Y, T).$$

with a velocity field $(K\cos\frac{\pi y}{2L}, 0)$

One looks for (steady) solution of the form

$$T(x + V_0 t, y), Y(x + V_0 t, y)$$

(travelling wave solution).

$(V_0 + K\cos\frac{\pi y}{2L}, 0)$ is the gas velocity in the reference frame in which the flame is stationary. Beresticky-Larrouturou conjectured that for very non uniform flow (K large) then there are solutions with $V_0 < 0$.

Benkaldoun-Larrouturou showed this *numerically* (with the same method as above), i.e.

- in some region $(V_0 + K\cos\frac{\pi y}{2L} > 0)$ the velocity is oriented from fresh mixture to burnt gases
- in some other regions the velocity is oriented from burnt gases to fresh mixture (locally dominant diffusive effects).

Since then, H. Beresticky - B. Larrouturou - P.L. Lions have shown that such solutions exist.

Expected improvements include the simulation for compressible medium.

11.4 Optimum design problem in fluids mechanics

"Riblets" as a drag reduction device.

Reference : O. Pironneau et al. INRIA + AMD

It has been shown experimentally that the inclusion of "riblets" on the surface of a solid can decrease the drag. One consider the following simplified situation.

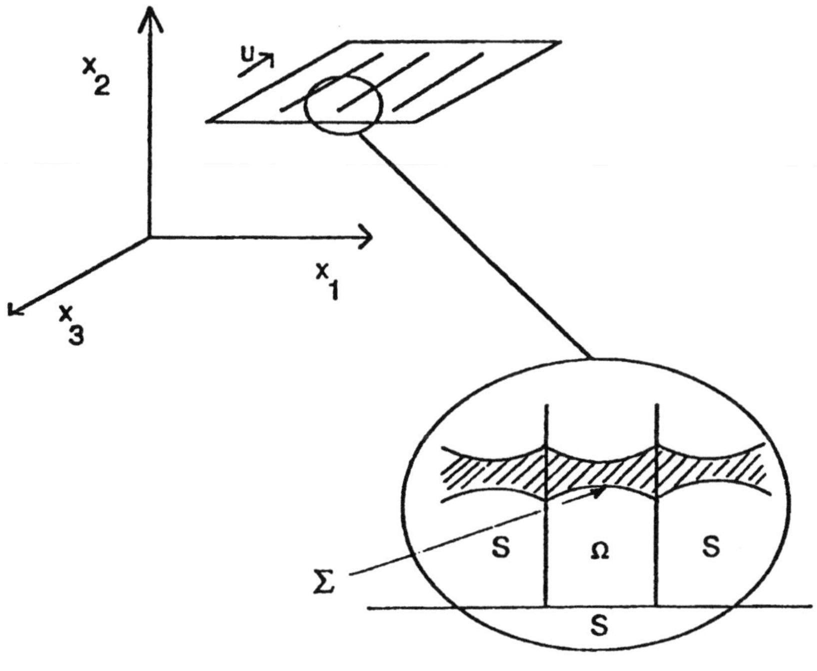

Figure 46.

Problem: find an optimal shape Σ.

The flow is described by Navier Stokes equations:

$$\begin{aligned} &u_t + u\nabla u + \nabla p - \nu\Delta u = 0 \\ &\nabla\cdot u = 0 && \Theta \times]0\,,T[\\ &u(x\,,0) = u^0(x)\,,\quad u(x\,,t) = u_\Gamma(x) && \partial\Theta \times]0\,,T[\end{aligned}$$

$\Theta = C - \Sigma$
$u_\sigma = 0$ on Σ

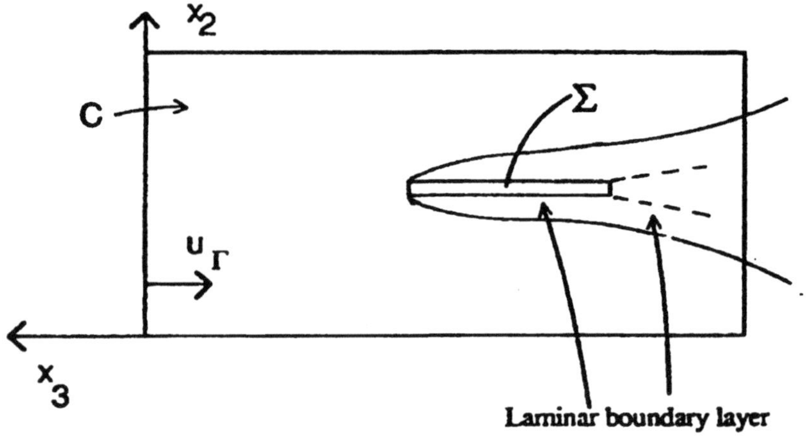

Figure 47.

One makes the following simplifying assumptions:

- stationary flow
- flow invariant by translation in x_3

$$u \; // \; x_3$$

- periodic riblets

One then reduces the model to *one cell.*

The horizontal component of the flow is then described by

$$(1) \qquad \begin{cases} -\Delta u = k \\ u|_\Sigma = 0, \quad \dfrac{\partial u}{\partial n}\Big|_S = 0. \end{cases}$$

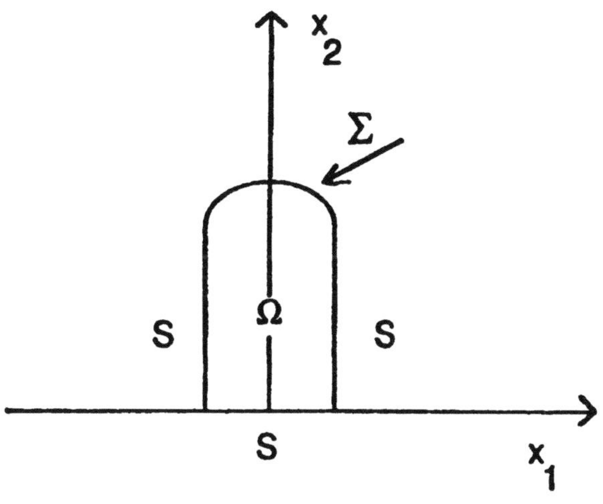

Figure 48.

Optimization problem : find an optimal shape of Σ in order to minimize the drag

$$D = -\frac{\nu}{2}\int_{\Sigma}\frac{\partial u}{\partial n}\,d\gamma \tag{2}$$

($d\gamma$ surface element of Σ.)

In order to have a meaningful problem, we consider the case of fixed total amount of flow

$$\int_{\Omega} u\,dx = d. \tag{3}$$

We then have the following equivalent formulations:

Formulation 1, for some set of admissible shape W

$$\min_{\Sigma\in W}\left\{-\int_{\Sigma}\frac{\partial u}{\partial n}\,d\gamma\,;\,k\,,\,u\text{ s.t. (1) (3)}\right\}$$

Formulation 2 (by Green's formula)

$$\min_{\Sigma \in W} \left\{ \frac{|\Omega|}{d} \int_{\Omega} |\nabla u|^2 \, dx \,, k \,, u \text{ s.t. (1) (3)} \right\}$$

Formulation 3 (taking $\upsilon = u/k$)

$$\max_{\Sigma \in W} \left\{ \frac{1}{|\Omega|} \int_{\Omega} |\nabla \upsilon|^2 \, dx \,, \upsilon \text{ s.t. (5)} \right\}$$

(5)
$$\begin{cases} -\Delta \upsilon = 1 \\ \upsilon|_{\Sigma} = 0 \,, \quad \dfrac{\partial \upsilon}{\partial n}\Big|_S = 0. \end{cases}$$

Admissible set:

- If there is no constraint, the problem has no solution
- By physical consideration, one deduces that there should be a relationship like

$$|\Omega| = f(\Sigma)$$

the simplest one is:

$$|\Omega| = a \text{ (constant)}.$$

The formulation 2 is then reduced to a known optimum design problem:

$$U_d(\Omega) = \left\{ u \in H^1(\Omega) \,, \quad u|_{\Sigma} = 0 \,, \quad \int_{\Omega} u \, dx = d \right\}$$

$$\min_{\Sigma \in W \,,\, u \in U_d(\Omega)} \int_{\Omega} |\nabla u|^2 \, dx$$

$$W = \{\Sigma / |\Omega| \geq a \,, \quad \Omega \subset \Theta\}$$

1) Mathematical Analysis

→ There exists at least one solution
→ in general, the flat shape is a local minimum
(in general, the semi-circle is a stationary point, not a local minimum).

2) Numerical study

Finite elements P^1 are used on the formulation 3 and give the discrete problem:

$$(6) \qquad \int_{\Omega_h} \nabla v_h \cdot \nabla w_h \, dx = \int_{\Omega_h} w_h \, dx \quad \forall w_h \in V_h$$

$$\max_{\Sigma_h \in W} \left\{ \frac{1}{|\Omega_h|} \int_{\Omega_h} |\nabla v_h|^2 \, dx \, , v_h \text{ solution of (6)} \right\}$$

and projected gradient method is used for the optimisation problem (with a parametrization of Σ).

Numerical results

- Flat shape is non optimal in general
- numerically obtained shapes (Figure 49) give 15% drag reduction (which appears as very high and could be a consequence of the simplified formulation).
- to be confirmed on *complete* Navier-Stokes analysis.

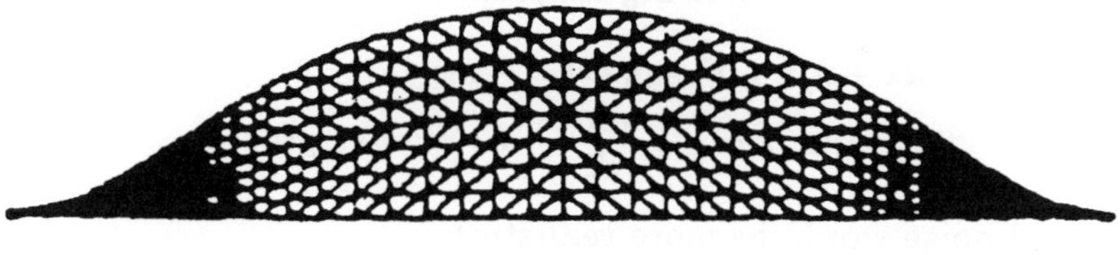

Figure 49.

11.5 Control of shock absorber

Ref. E. Pardoux et al. - INRIA/Renault

Active suspension systems are considered for trucks. The following simple situation is considered:

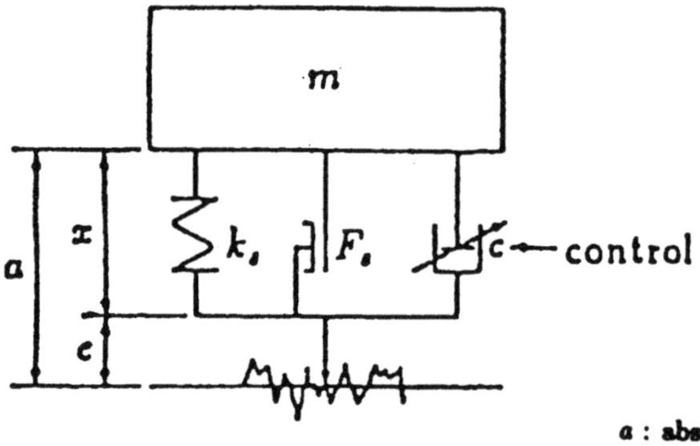

a : absolute deplacement
x : relative deplacement
e : random input (road profile)
m : mass
c : variable damping
k_s : stiffness coefficient
F_s : friction coefficient

Figure 50.

Model

$$m\ddot{x} + c\dot{x} + k_s x + z = -m\ddot{e}$$
$$\dot{z} + \beta(|\dot{x}|z + \dot{x}|z|) = (k - k_s)\dot{x}$$
$$T_r\dot{c} + c = u$$

Simplified model : the following simplified model is used

$$m\ddot{x} + u\dot{x} + k_s x + F_s \operatorname{sgn}(\dot{x}) = -m\ddot{e}$$

random input $\ddot{e} = -\sigma \dfrac{dw}{dt}$ $\quad w$: Wiener process.

[finite band noise would be more realistic]

Control problem : one rewrites the equations as

$$X = \begin{pmatrix} x \\ \dot{x} \end{pmatrix} \quad dX_t = b(x_t)dt + \begin{pmatrix} 0 \\ \sigma \end{pmatrix} dw_t$$
$$b_1 = \dot{x}\,,$$
$$b_2(u\,,x) = -\frac{1}{m}(u\dot{x} + k_s x + F_s \operatorname{sgn}(\dot{x}))$$

minimize $J(u)$:

$$J(u) = \lim_{T\uparrow+\infty} \frac{1}{T} \int_0^T f(u, X)dt,$$

$$f(u, X) = |b_2(u(X), X)|^2, \quad 0 \leq u(X) \leq \bar{u}.$$

(norm of absolute acceleration).

This is a relatively well-known situation in stochastic control and can be analyzed through H.J.B. equation

$$L(u)V(X) \equiv b(X)\cdot \frac{\partial V}{\partial X} + \frac{\sigma^2}{2} \frac{\partial^2 V}{\partial X_2^2}$$

$$\inf_w (L(u)V(X) + f(u, X)) = \lambda$$

where λ will be the optimal time average cost.

Discrete approximation

Finite differences giving a control problem for finite state Markov process is used (for a given control law $u(X)$).

On a bounded domain:

$$D = [x_{\min}, x_{\max}] \times [\dot{x}_{\min}, \dot{x}_{\max}]$$

(with *reflecting* boundary conditions).

The corresponding H.J.B. equation is

$$\min_{0 \leq u_i \leq \bar{u}} [L_h(u)v + f(u)]_i = \lambda_h \quad i = 1, \ldots, d$$

which can be solved by policy iteration method.

The main point is to be able to test *simple strategies*, i.e. suboptimal control: for instance

$C1$ – minimization of $f(u, X)$
$C2$ – parametric optimization :
$u_\theta(X) = (\theta_1 + \theta_2 x \operatorname{sgn}(\dot{x}))^+$

θ being such that $J(\theta) = J(u_\theta)$ is minimized.

Numerical experiments show good performance of the suboptimal control $C2$ including on the complete model.

11.6 Identification and monitoring of mechanical structures.

Ref. A Benveniste et al. - INRIA + IFREMER

One starts with the following model of a structure like the one in Figure 53 (M, C, K are matrices)

$$M\ddot{Z} + C\dot{Z} + KZ = E$$
$$Y = LZ \quad \text{observation.}$$

A discrete time model is then:

$$X_{t+1} = FX_t + V_{t+1}, \quad V_t \text{ white noise}$$
$$Y_t = HX_t$$
$$Q_t = \text{cov}\,(V_t) \text{ \textit{nonstationary}}, \text{ and unknown.}$$

It is known that one can write an ARMA representation

$$Y_t = \sum_{i=1}^{p} A_i Y_{t-i} + \sum_{j=0}^{p-1} B_j(t)\varepsilon_{t-j}$$

V_t is involved only in $B_j(t)$

ε_t standard white noise

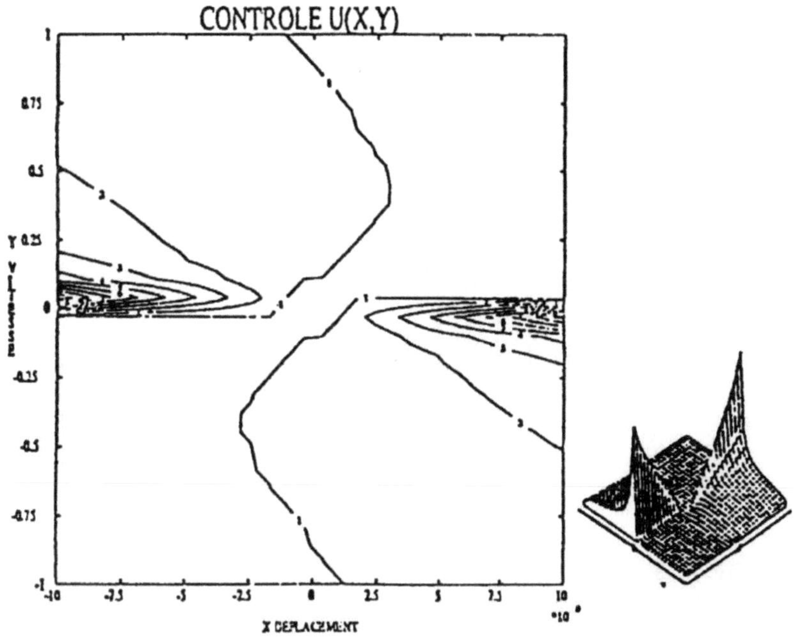

Figure 51. optimal control (λ = 2.22)

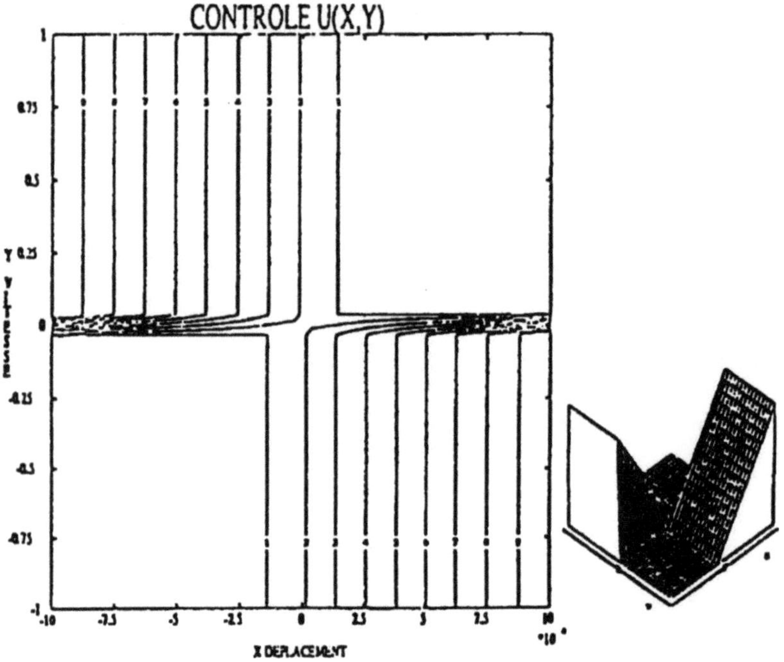

Figure 52. suboptimal control C2 - (λ = 2.37)

The problem is to identify modal characteristics of the structure.

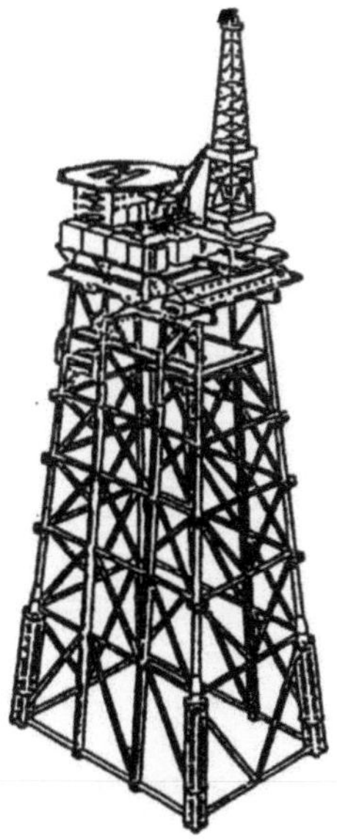

Figure 53.

The identification in order to be able to make a modal analysis of the system is reduced to the estimation of $(A_i)_{i=1,p}$.

The identification method uses the orthogonality of

$$W_t = Y_t - \sum_{i=1}^{p} A_i Y_{t-i}$$
$$\text{w.r.t. } Y_{t-p},\ Y_{t-p-1} \cdots$$

The only necessary information is the set of observation covariances

$$R_m = E(Y_{t+m} Y_t^T)$$

replaced, in practice, by estimates

$$\hat{R}_m = \sum_{t=1}^{s-m} Y_{t+m} Y_t^T$$

This is more or less an "instrumental variable" method, which can be shown to be consistent *even* with $B_j(t)$ are not constant.

Monitoring : the problem is now the following:

Given $\theta_0^T = (A_p^0, \ldots, A_1^0)$ and measurements $Y_1 \ldots Y_s$	is θ_0 still a good description ?

For that purpose, one defines a statistics:

$$\chi(\theta_0, D, (Y_t)_{1 \le t \le s})$$

D : assumed changes of θ ($\delta\theta \in D$)

and uses a decision rule like:

$$\chi < \text{Treshold} \Rightarrow \delta\theta = 0$$
$$\chi \geq \text{Treshold} \Rightarrow \delta\theta \neq 0, \quad \delta\theta \in D$$

The statistic χ is obtained by the following scheme. Taking

$$U_N^T(s) = (A_p^0 \ldots A_1^0 - I)\, \mathrm{H}_{p+1,N}(s)$$

$\mathrm{H}_{p+1,N}$ Hankel matrix formed with the covariances of observation

$$(\mathrm{H}_{p+1,N})_{ij} = R_{i+j-1}$$
$$R_m = E(Y_{t+m} \cdot Y_t^t)$$

One can show that

$$\text{any change } \delta\theta \text{ on } \theta \rightarrow \delta U_N \underset{\text{(in average)}}{=} \mathrm{H}\delta\theta$$

and $U_N(s)_{s \to \infty} \rightarrow$ Gaussian.

In practice, one takes $N = p$ and

$$\chi = U_N^T \hat{\Sigma}_N^{-1} U_N$$

$\hat{\Sigma}_N$ estimate of the covariance of U_N

Several improvements are of interest:

- variable location of sensors
- monitoring of a group of modes,

- implementation of fast algorithms...

11.7 Concluding remarks

1. the wide spectrum of problems is obvious

2. an essential problem is *modelling* which is in fact a major requirement of industrial problems

3. development of integrated tools is believed to be a major factor of improvement of use of math in industry.

Examples of integrated tools are often seen in mechanics for instance and we think that such developments are of essential interest in control as well. An example is provided by the Pandore system developed by J.P. Quadrat's team at INRIA (the main functions of Pandore are shown on Figure 54)

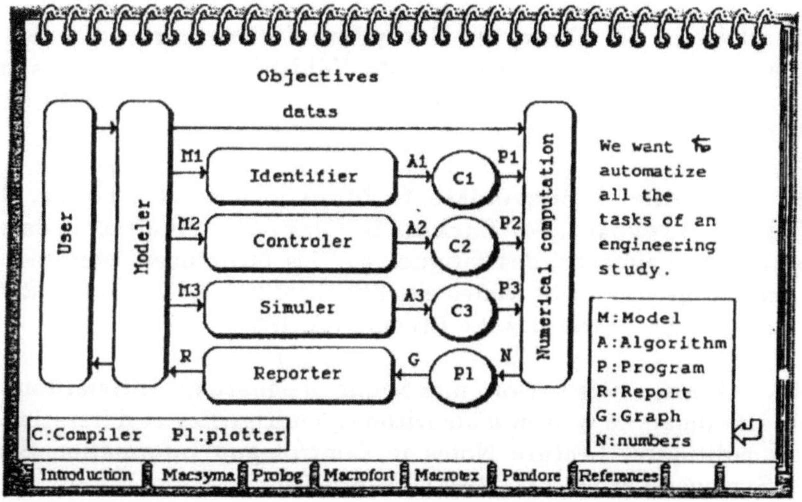

Figure 54.

11.8 References

[1] **INRIA**, activity report 1987.

[2] **F. Benkaldoun - B. Larrouturou** , Adaptive calculation of wrinkled flames, Sixth international symposium on finite element methods in flow problems, Antibes, (1986), (actes to appear).

[3] **F. Benkaldoun - B. Larrouturou** , Simulation numérique adaptative de flammes plissées, Colloque national d'analyse numérique, Port-Barcarès, (1986).

[4] **G. Arumugam, O. Pironneau** , On the problem of riblets as a drag reduction device, to appear in Optimal control application and methods, ed. Bion El Pierson, John Wiley & Sons, Chichester, U.K.

[5] **J.P.Chancelier, C. Gomez, J.P. Quadrat, A. Sulem** , Un système expert pour l'optimisation de systèmes dynamiques. Congrès INRIA d'Analyse Numérique, Décembre 1985.

[6] **S. Bellizzi, R. Bouc, F. Campillo, E. Pardoux** , Contrôle Optimal Semi-Actif de Suspension de Véhicule, Lecture Notes in Control and Information Sciences, Springe'er-Verlag, 111, (1988), pp. 689-699.

[7] **M. Basseville, A Benvéniste, G. Moustakides, A. Rougée, G. Le Vey, M. Prevosto, M. Barkin, B. Olagnon** , Analyse modale, surveillance, diagnostic des fatigues sur les structures soumises à une excitation naturelle. Actes de l'Ecole d'Eté CEA/EDF/INRIA sur l'Analyse des vibrations. Le Bréau, Juin 1986.

[8] **M. Akian** , Résolution numérique d'équations d'Hamilton - Jacobi - Bellman au moyen d'algorithmes multigrilles et d'itérations sur les politiques, Lecture Notes in Control and Information Sciences, Springer-Verlag, 111, (1988), pp. 629-640.

12. Modelling and Optimizing Fuel Consumption and Pollutant Emissions in an Alfa Romeo Car

Emilio Spedicato

Dipartimento di Matematica

Istituto Universitario di Bergamo

via Salvecchio 19, 24100 Bergamo

12.1 Summary

Using data from an Alfa Romeo car, functions are constructed that describe the rate of fuel consumption and pollutant (HC, CO, NO_x) emissions. The total fuel consumption over a European cycle is minimized under drivebility constraints and EEC limits on the emissions of the pollutants. The found solution gives a significant improvement over the previously optimized performance. Robustness of the model and stability of the solution is indicated by sensitivity analysis.

12.2 Introduction

Among the tasks that presently face the automotive industry of great importance is the one of reducing fuel consumption (in view of energy conservation policies) and of keeping pollutant emissions as low as possible (in view of environmental and health considerations). There are many ways of dealing with these problems, including better aerodynamical design, use of new materials and new fuels, innovative concepts of engine (for instance hybrid thermo-electrical engines) and so on. In this paper we shall only consider how the use of mathematical techniques can make the electronic control of certain engine parameters more efficient. We shall assume that the car is provided with sensors, which measure certain state parameters and input the measured data to a microcomputer. The microcomputer elaborates the data according to the criteria described further on in order to determine optimal values of certain control parameters, to be inputed to suitable actuators. Feedback action to take into account delays and errors is not considered here, being of course of great importance in practice.

The described research has been performed in a period of four years, in the framework of the Progetto Finalizzato Trasporti of the National Science Council of Italy (CNR) and in strict collaboration with Alfa Romeo, which has provided data relating to a 2000 cubic centimeter engine. Further developments are expected with the introduction of additional state and control parameters. Commercial applications are expected in the nineties.

12.3 The mathematical problem of optimal electronic engine control

The multiobjective problem of reducing fuel consumption and pollutant emissions has been set by the automotive industry as the following single object optimization problem. Let the engine be characterized at time t by a set of state parameters $s = s(t)$ and a set of control parameters $u = u(t)$. Let $f(s(t), u(t), t)$ be the rate of fuel consumption at time t; note that the variable t appears as an independent argument, because f is not a univariate function of s and u, its value depending also on the trajectory through which the point $s(t), u(t)$ is reached. Let similarly $f_i(s(t), u(t), t)$ be the rates of emission of certain I pollutants, $i = 1, \dots, I$. Let $Q = Q(s(t), t)$ be the region of implementable values for the control $u(t)$. Let s be a given trajectory of duration T in the space (s,t). Let c_i, $i = 1, \dots, I$, be the maximum admitted values of the total emissions of the i-th pollutant along S. Then we have to solve the following
Problem P1:

$$\underset{u(t)}{\text{minimize}} \qquad F = \int_0^T f(s(t), u(t), t)dt$$

subject to: $s(t) \in S$; $u(t) \in Q$

$$\int_0^T f_i(s(t), u(t), t)dt \leq c_i \qquad i = 1, \dots, I$$

Problem P1 in practice is simplified as follows:

- the history dependence in f and f_i is neglected, say f and f_i are assumed to be functions only of $s(t)$ and $u(t)$

- the region Q is given in the form of lower and upper bounds for $u(t)$, say $u_1(t) \leq u(t) \leq u_2(t)$

- the integrals are approximated by a summation over a finite number N of points along the cycle, so that instead of evaluating the function $u(t)$ for $0 \leq t \leq T$ the problem is reduced to evaluate the N controls $u(t_1), \dots, u(t_N)$.

Therefore we obtain the following simplified Problem P2:

$$\underset{u(t_1)\dots u(t_n)}{\text{minimize}} \; F = \sum_{j=1}^{N} f\,(s(t_j), u(t_j))dt_j$$

subject to: $s(t_j) \in S$

$$u_1(t_j) \leq u(t_j) \leq u_2(t_j)$$

$$\sum_{j=1}^{N} f_i\,(s(t_j), u(t_j))dt_j \leq c_i$$

$$(\, i = 1, \dots, I; \; j = 1, \dots, N)$$

If the control u is a vector of dimension q, then Problem P2 is a mathematical programming problem of n = Nq variables and 2Nq + I constraints, once the cycle S is given and assuming that the functions f, f_i are known.

In the case that we have studied, with data relating to the Alfa Romeo engine, both the state and control parameters are two dimensional vectors, consisting the first of the speed of rotation and the torque, and the second of the spark advance angle and the air/fuel ratio. The cycle S is assumed to be the European cycle, a trajectory in the space of the state variables that attempts to simulate the average behaviour of a European car (there exist similar American and Japanese cycles). Of the many pollutants emitted by a car, only three have been considered, following current EEC legislation, say *HC* (which is cancerogenus), *CO* (dangerous to the respiratory system) and NO_x (a major culprit of acid rain inter alia). The limits c_i are those fixed presently by EEC legislation. The cycle has been discretized in N = 16 discretization points (working points), chosen, with the value of the weights dt_i, with engineering criteria. For an analysis of the discretization effects on the European cycle see Rudzynska and Spedicato [1].

12.4 Construction of the engine functions

The functions f and fi are not known in explicit form and they are not easily computable by solving suitable dynamical equations. In practice they are defined or estimated through data collected in laboratory. There the engine is set running at a certain point s_j in the space of the state variables. Then several values of the control parameters are implemented, defining a more or less regular grid in the control space. The values of the functions f, f_i are then measured.

To reduce history effects, the measurements are made after warming up of the engine. The results of the experiments is the availability of a two dimensional table (in our case consisting of about one hundred data) for each function and each working point.

Once the tables are obtained, there are two possible approaches to Problem P2. One is to select the best sequence of implemented controls, say the one that minimizes F subject to the constraints. This approach can be realized using discrete optimization techniques. The other one is to use the data to construct a continuous mapping of the functions and then apply continuous optimization techniques.

This is the approach that has been mainly followed in the recent literature and that we too have adopted.

Most researchers have used linear regression techniques to estimate the functions. For instance Powell et al. [2] have used the old well known BMD package for linear regression (using Legendre polynomials). Due to the availability since 1982 of the very efficient package DASL developed at the Physical National Laboratory in Teddington, UK, for one and two dimensional data fitting by Chebichev polynomials or splines, we have decided to make use of it. After extensive testing the conclusion was that fitting by splines was preferable, since approximation of the data of comparable quality could be obtained with splines using lower degree polynomials. The degree and the knot location of the splines were determined on heuristic criteria, since statistically based automatic criteria were not available. Incidentally the use of the splines was instrumental in discovering some serious problems in the first given set of data (outliers and irregular working of the engine in certain parameters areas); quite remarkably such problems had gone unnoticed in an independent fitting of the data by another analyst team using linear regression techniques. In Table 1 we give the used degree of splines and the average relative error in the fitted Alfa Romeo data.

12.5 Computing optimal control values

Problem P2, with f, fi obtained by a smooth fitting of the data, constitutes a nonlinear programming problem, of moderately large size in our case (the number n of variables is 32 and the number p of inequality constraints is 67). This problem can be solved straightforwardly using existing optimization packages. If more variables were included (in future research we plan to include exhaust gas recycle ratio and continuous transmission ratio) or more points along the cycle were considered, or also more refined definition of the drivebility region Q were given, then the increase in the dimension would make convenient to exploit the structure of the problem to reduce the computational cost. A possible way of taking into account the structure is the following.

Consider the first order Kuhn-Tucker conditions of problem P2:

$$(1) \qquad \nabla_{u_j} f(s_j, u_j) + \sum_{i=1}^{3} \lambda_i \nabla_{u_j} f_i (s_j, u_j)\, dt_j + \mu_{j1} - \mu_{j2} = 0$$

$$(2) \qquad \lambda_i \left(\sum_{j=1}^{N} f_i (s_j, u_j)\, dt_j - c_i \right) = 0$$

$$(3) \qquad \mu_{j2} (u_j - u_{2j}) = 0$$

$$(4) \qquad \mu_{j1} (u_{1j} - u_j) = 0$$

$$(5) \qquad \lambda_i \geq 0$$

$$(6) \qquad \mu_{j1}, \mu_{j2} \geq 0$$

where $i = 1, 2, 3,\ j = 1, \ldots, N,\ s_j = s(t_j),\ u_j = u(t_j),\ u_{1j} = u_1(t_j),\ u_{2j} = u_2(t_j)$ and $\lambda_i, \mu_{j1}, \mu_{j2}$ are the Lagrange multipliers associated respectively with the total pollutant emissions and the lower and upper bounds in the control parameters. Inspection of the above conditions shows that the control parameters at different discretization points are coupled only through the multipliers λ_i. If these multipliers were known, the controls at the j-th working point could be obtained independently by dropping conditions [2] and [5] (under the usual function conditions which guarantee unicity of the Lagrange multipliers). Observe that the remaining equations are just the Kuhn-Tucker conditions of the following Problem P3:

$$\underset{u_j}{\text{minimize}} \quad F = \left(f(s_j, u_j) + \sum_{i=1}^{3} \lambda_i f_i (s_j, u_j) \right) dt_j$$

subject to: $u_{1j} \leq u_j \leq u_{2j}$

Note also that in such a case the weight dt_j does not influence the solution, so that it can be actually removed from the definition of F. The considered use of the structure implies that N problems of dimension two (in our case). have to be solved, instead of a single problem of dimension 2N. This results generally in a reduction in the computational cost.

The considered simplification has been adopted in the past (see for instance Powell et al.) [2], when efficient software was not available. The initial estimate of the λ_i's was set to zero (this being equivalent to putting no limits on the pollutant emissions) and the multipliers were then increased according to some heuristics until the total amount of the emissions, computed once the optimization was performed, were below the legal limits. More sophisticated ways of updating the multipliers could clearly be devised in the framework of the augmented Lagrange multipliers or recursive quadratic programming algorithms used for the optimization.

After experimenting with different optimization packages, our final choice was in favour of the code OPRQP in the package OPTIMA developed at the Numerical Optimization Centre of the Hatfield Polytecnic, UK, and based upon the recursive quadratic programming algorithm of Biggs [3]. The code had performed very well in the experiments of Schittkovsky [4] and has confirmed its robustness and efficiency (coupled with simplicity of use and remarkably short length of the program) in the case under consideration. The code was run in single precision on the Gould computer at Alfa Romeo, producing a solution generally in about half a minute for the several tried starting points. We considered eleven starting points generated as follows. One consisted of the optimal point known to Alfa Romeo and presently implemented on the engine. Ten more points were obtained by randomly generating $3 \cdot 10^5$ points (a task which took a few hours of CPU time) inside the drivebility region and memorizing, among the points that satisfied also the limits on the emissions, those which gave the least fuel consumption. Quite remarkably, none of the points generated in this way, gave a lower fuel consumption than the Alfa Romeo point, a result clearly representative of the "dimensionality curse" that affects random algorithms for optimization.

From all the considered starting points the algorithm has found the same solution giving, with respect to the previous best point known to Alfa Romeo, a 6% reduction in fuel consumption, a 20% reduction in HC, a 55% reduction in CO, but a 20% increase in NO_x. In Table 2 we give the following information: initial value of consumption and emission for the first five starting points (P1 is the Alfa Romeo point , values in the computed optimal solution (point PF) and legal EEC limits (last column). While it is not possible to claim that the problem has a unique solution, the obtained convergence to a same point is in agreement with previous findings of Powell, private communication). We should however remark that when OPRQP was applied to the problem with the functions evaluated from the first set of data (which, as as said before, had irregularities detected through the use of the splines) different starting points produced different solutions!

In the determined solution no constraint is active. By sufficiently reducing the limits on the pollutants, active constraints appear in both the emission and drivebility constraints. For instance when the legal limit for NO_x is reduced from 18,21 to about 10,60 it becomes active. Similarly HC becomes active when it is reduced from 12,80 to about 2,90 (all other limits left unchanged). If the limit for NO_x is further reduced .to about 9,90, no solution is found close to the solution corresponding to a slightly higher values of the limit (but a solution with higher fuel consumption exists in another region). This fact seems to indicate that with more stringent emission policies multimodality and discontinuity problems are expected.

Multimodality is also expected when additional control parameters are included (see Radtke et al. [5] for the existence of two solutions when transmission ratio is included). Once the optimal solution was obtained, we applied sensitivity analysis techniques to evaluate its stability and the robustness of the model. First Montecarlo techniques were used to evaluate the effect of perturbation in the data and in the solution. Randomly generated errors with Gaussian distribution, zero mean and different variances were added to the data to simulate variability among cars. The perturbed data were then fitted with the same set of splines previously determined and the corresponding optimal control parameters were recomputed. Their expected variation was evaluated by averaging over a large set (three hundred) of perturbed data. The results indicate that the model is robust, stronger variations appearing only in the expected value of CO. We consider the high sensitivity of CO a possible consequence of the poor quality of the Alfa Romeo data for CO, as clearly appears from Table 1 (actually a closer inspection of the data shows the presence of a physically unaccettable peak for large values of the rotation speed).

Random variations in the optimal control parameters were then introduced in similar way, by treating the two controls either separately or together. The expected variations in fuel consumption and emissions were again evaluated by averaging on a large number (four hundred) of cases. The solution appears to be stable from the point of view of consumption and emissions, again with a higher sensitivity of *CO* (its variation being one order larger). An additional interesting result was obtained when the sensitivity of our solution was compared with that of the previous best Alfa Romeo point, which turned out to be generally more sensitive by a factor two. Overall it appears that an improvement over the old policy can still be obtained in presence of errors up to one per cent in the actuators (errors well below the technologically and commercially feasible limits). The results of sensitivity analysis by the described Montecarlo technique have been confirmed also by using deterministic techniques as those discussed by McKeown [6]; see Finardi [7] for details.

Finally an estimate of the optimal control parameters over the whole engine working region has been obtained by fitting the optimal controls in the discretized European cycle with orthogonal polynomials and interpolating splines, see Spedicato and Vespucci [8]. The fitting indicates that the control surface is smooth.

12.6 Conclusions

In this paper we have synthesized the work done at the University of Bergamo in collaboration with Alfa Romeo on the optimal control of engine parameters (this work is fully documented in several technical reports, see for instance [9] [10] [11] [12] [13]). The results have indicated that significant improvements can be obtained by using the described techniques. In the next future we expect to apply these techniques to more control variables. Also the development of on line model building and optimization is forecast, making use of the greater power offered by parallel processors. For preliminary results in this directions, see Dixon and Spedicato [14], [15].

12.7 References

[1] Rudzynska K. and Spedicato E., Optimal engine control discretization effects on fuel consumption and pollutant emissions, Quaderno DMSIA 87/3, University of Bergamo, 1987

[2] Powell J., Glaser I., Randall K. and Hosey R.: Closed loop control of automotive engines, Final report for the Transportation System Center, Department of Aeronautics, Stanford University, 1980

[3] Biggs H.C.B.: Recursive quadratic programming based on penalty functions for constrained minimization, in Nonlinear Optimization: theory and algorithms, Dixon L.C.W., Spedicato E. and Szego G.P. editors. Birkhauser, Boston, 1980

[4] Schittkovsky K.: A numerical comparison of optimization software using randomly generated test problems, Report 45, Institute of applied mathematics, University of Wuerzburg, 1978

[5] Radtke R. and Ennerweher L.: Optimization of a continuosly variable transmission with emission constraints, SAE Report 107, 1981

[6] Mc Keown J.J.: Sensitivity analysis of nonlinear programming problems, in Nonlinear optimization: theory and algorithms, see above

[7] Finardi M.: Algoritmi per analisi di sensitivita' ed applicazione al controllo elettronico delle auto, Dissertation, Computer Science Department, University of Milano, 1987

[8] Spedicato E. and Vespucci M.T.: Mapping optimal engine control parameters, Technical Report, PF Trasporti, University of Bergamo, 1987

[9] Abaffy J., Muscillo C., Spedicato E., Vespucci M.T.: Una procedura per il fitting mediante funzioni splines di dati sperimentali, relativi a funzioni motoristiche, Technical Report, PF Trasporti, University of Bergamo, 1987

[10] Vespucci M.T.: Approssimazione di funzioni motoristiche mediante splines bidimensionali, Technical Report, PF Trasporti, University of Bergamo, 1987

[11] Abaffy J., Muscillo C., Spedicato E. and Vespucci M.T.: Un codice per il controllo ottimale di parametri motoristici, Technical Report, PF Trasporti, University of Bergamo, 1987

[12] Spedicato E. and Vespucci M.T.: Ottimizzazione del consumo di carburante di un'auto a benzina su ciclo Europa con vincoli sull'emissione di inquinanti, Technical Report, PF Trasporti, University of Bergamo, 1987

[13] Spedicato E. and Vespucci M.T.: Analisi di sensitivita' di un modello ottimizzato di consumo carburante in auto a benzina, Technical Report, University of Bergamo, 1987

[14] Dixon L.C.W. and Spedicato E. : Software developments for numerical on-line parallel optimization of engine fuel consumption 1: a survey of the problem, Report NOC 149, Hatfield Polytechnic, 1984

[15] Dixon L.C.W. and Spedicato E.: Software developments for numerical on line parallel optimization of engine fuel consumption II: state of the art, Report NOC 172, Hatfield Polytechnic, 1985.

Table 1: data on spline fitting

Point	Spline degree Fuel	HC	CO	NOx	Number of Fuel	HC	CO	NOx	Average relative error Fuel	HC	CO	NOx
1	3	3	3	3	2	3	3	3	.76	1.3	17.6	1.5
2	3	3	3	3	1	1	3	3	.24	.86	.68	.48
3	3	3	3	3	1	2	3	2	.32	.43	1.3	1.2
4	3	3	3	3	2	2	2	3	.54	2.3	6.1	2.2
5	3	3	3	3	1	1	2	2	.32	.36	18.9	1.9
6	3	3	3	3	1	1	2	1	.13	.43	178.3	10.3
7	3	3	3	3	1	2	2	2	.14	.32	73.9	3.7
8	3	3	3	3	1	2	1	2	.28	1.3	52.5	22.1
9	3	3	4	3	2	2	1	3	.43	1.0	64.1	1.7
10	3	3	3	3	2	2	1	3	.20	.64	31.6	1.2
11	3	3	3	3	0	1	1	2	.86	2.5	99.9	3.4
12	3	3	3	3	0	2	1	3	.52	1.3	48.8	2.1
13	3	3	3	3	1	2	1	3	.22	.16	71.9	.31
14	3	3	3	3	1	1	1	2	.15	.51	38.1	1.0
15	3	3	3	3	1	1	1	2	.31	1.0	70.1	1.9
16	3	3	3	3	1	1	2	2	.28	2.2	38.3	2.7

Table 2: starting point and final point function values.

Funct.	P1	P2	P3	P4	P5	PF	Bounds
Fuel	275.3	275.8	276.2	276.2	276.2	252.5	
HC	3.37	3.1	3.51	3.30	3.48	2.93	12.68
CO	28.08	23.4	41.92	31.17	44.65	10.06	155.12
NOx	8.15	8.2	8.63	7.8	8.1	10.65	18.21

Note: The units are grams per hour and kilowatt

13. The Role of SIAM in Promoting Industrial Mathematics

Prof. I. Stakgold

President Elect of SIAM

(Society for Industrial and Applied Mathematics)

University of Delaware, USA

My talk is entitled SIAM's role in promoting industrial mathematics. This is an improperly posed problem since the term industrial mathematics is not well defined. Applied mathematicians in industry do many different things. Some are specialists, some are general consultants within the organization, some are members of team projects, and, of course applied mathematicians may move from one role to another during their careers. Consulting work and project work are probably the distinguishing features of industrial mathematics. What such applied mathematicians publish in scientific journals is, however, much more academic-style applied mathematics, well polished with most of the distinctive industrial part left out. More about that later, but first let me give you some general information, including a little history, about SIAM.

The Society was founded in 1952 and celebrated its 35th anniversary last year. Why was the Society founded? During the second World War and in the immediate Post War period we saw a surge in applied mathematics and the emergence of machine computation. Operations Research also became a discipline at that time.

Those who worked in these areas came from mathematics, physics and engineering. Their needs were not met by existing scientific societies. The American Mathematical Society, for instance, did not show much interest in developing programs and activities for applied and computational mathematicians.

These circumstances led to the establishment of SIAM. Other Societies such as the Operations Research Society of America and the

Association for Computing Machinery were also founded at around the same time. The character and personality of SIAM developed somewhat later in the sixties. The core of the membership was in physical applied mathematics and numerical analysis. Among the Presidents of SIAM in that period were Joseph LaSalle, Alston Householder, J. Barkley Rosser, Wallace Givens, Herb Keller and C. C. Lin. George Carrier and Joseph Keller also held positions with the Society.

Where does SIAM stand today? It is clear that we are now covering more fields than ever before as you will see as I describe our services.

The main services provided by SIAM for its members are publications and meetings. We publish 10 scientific journals: The SIAM journals of Applied Math., of Math. Analysis, of Computing, of Control and Optimization, of Numerical Analysis; The SIAM Review; Theory of Probability; and the latest 3, the SIAM Journals of Scientific and Statistical Computing, of Matrix Analysis and applications, and of Discrete Mathematics.

This is perhaps our greatest contribution to the community, but we also have a book program which is being expanded and we publish SIAM News, a monthly information bulletin which has a wide circulation.

Our other principal activity is meetings: The Annual national meeting, special topical conferences and workshops. Some of these are organized by our special interest groups (Optimization, Control, Linear Algebra, Discrete Mathematics, Supercomputing, and the most recent one, Dynamical Systems). These groups reach out to other organizations with overlapping interests. Some of these conferences have been among our most successful. On the international scene we cooperate with GAMM in Germany, IMA in Britain, and the newly-formed SMAI in France. For instance, we are cosponsoring with SMAI and INRIA a conference on computational combustion in 1989. There are also prospective applied mathematics societies in Italy, Spain, Benelux, Japan, and China. Scandinavian applied mathematicians have chosen instead to become a section of SIAM, the Nordic section. A particularly important project in which we, together with our colleagues throughout the worlds are involved is the ICIAM 91 meeting International conference in industrial and applied mathematics to be held in Washington from July 7 to July 12, 1991. This is the second such conference. The first, which was held in Paris in 1987 was very successful. I hope to see many of you there.

SIAM is also concerned with education, an area in which I believe we should make greater efforts. SIAM also makes various awards (some jointly with other societies). The Von Neumann lecture is our most prestigious award and is delivered at the annual meeting. Jacques-Louis Lions received this award in 1986 and gave a brilliant lecture that has been published in the SIAM Review of March 88.

Now let me turn to SIAM's relationship to mathematicians in industry and government. I believe our Society serves its industrial and government members quite well. It must be admitted, however, that these industrial mathematicians are in some ways close in outlook to our academic members. After all, IBM Research, Bell Labs, the National Bureau of Standards, Los Alamos, are much like a university without the commitment to teaching. Moreover, American professors must compete for government grants. A colleague at IBM has said that he would never leave the peaceful environment of IBM for the tense and competitive world of the university. If we look at nonacademic articles in the SIAM J. of A. M., most are by scientists at large research laboratories of the type already mentioned. What they are publishing in our journals is not precisely the industrial mathematics that stems from industrial problems. This brings me to an important question: How to publish industrial mathematics. The subject does not lend itself readily to orthodox publication channels. The process leading to the solution (or perhaps solution is the wrong word - clarification, formulation might be better) is the essential element, rather than the finished product. What distinguishes industrial mathematics from its academic counterpart is the give and take between engineer and mathematician, the struggle with the modeling aspects, the choice of good analytical and numerical approaches (in the industrial context, the best is often the enemy of the good: il meglio e' spesso il nemico del bene).

There are not many published articles that convey the flavor of this process. Our European colleagues may be more successful in disseminating their philosophy through organs such as the IMA journals, especially under the editorship of John Ockendon.

By the way, teaching industrial mathematics is equally difficult. Two of the best programs in the U.S., the ones at Claremont and RPI will be discussed at this workshop.

Returning to SIAM's contributions, I should point out that mathematical modeling is often a part of the classroom notes in SIAM Review. SIAM News has articles on industrial problems. On the educational front we sponsor a student competition on mathematical modeling with an award to the winner. Our university student - faculty chapters often have programs of talks by industrial and government mathematicians. Some of our SIAM sections such as the Great Lakes section have considerable industrial government-academic interaction. We also run workshops which have an industrial component. Last year, we organized an international workshop at Claremont on industrial mathematics.

This is a modest list of programs, but we hope to achieve more in the next few years. The educational part is vital. We need people to meet the industrial demand. As Ed David, Former President of Exxon Research and Engineering has said: Today's high technology is a mathematical technology. As you know, the number of advanced degrees in mathematics in the United States has dropped considerably in recent years despite the market needs. Perhaps it is because our programs are not sufficiently interesting. Innovative ideas in industrial mathematics education do a great deal to recruit applied mathematicians into the profession. I believe we have much to learn from schools such as SASIAM here in Valenzano, and various other programs in Europe which provide an alternative to the standard applied mathematics programs.

Now let me say something about SIAM's plans for the next few years. Some of these projects are underway, others are being considered.

1 - The Math Mobile.

For elementary, middle and high schools. We want to develop an innovative mobile educational resource focused on mathematics and its applications. The Math Mobile is a take-off from the Book Mobile which is a way of bringing books to rural communities. The Math Mobile will be designed to provide students with a variety of exhibits, videotapes, computer displays and hands-on activities appropriate to the student's age and mathematical readiness. Resources for teacher education will be provided to build upon and reinforce the Math Mobile experience in the classroom. This is propaganda for applied and industrial mathematics that will emphasize the relevance of mathematics to the real world.

2 - Survey of Master's programs in Applied Mathematics.

This is being done by SIAM in conjunction with the MS2000 project, an assessment of collegiate mathematical resources in the U.S. We hope to describe some model programs.

3 - Informational Exchange.

We hope to work with ECMI (European Consortium for Mathematics in Industry) and similar organizations to build a collection of useful materials, modules and information on industrial mathematics and on industrial-academic interaction.

4 - Workshop on industrial mathematics.

We want to continue with the kind of workshop that was started at Claremont. We are also trying to increase industrial participation in our regular conference program. Three conferences scheduled for 1989 and 1990 are particularly well suited for this purpose.

a) Mathematical and computational issues in geophysical fluid and solid mechanics.

b) Geometric design.

c) Simulation of computer systems.

5 - Visiting lecturer program.

We are setting up a roster of industrial and government mathematicians who are willing to give presentations at our colleges.

6 - Internship programs.

We want to help universities develop internship programs for their graduate students in applied mathematics. Such programs are very successful in statistics. Many universities also have successful statistical consulting laboratories. What is needed to develop similar structures for applied mathematics? It is perhaps a bit ironic that I had to come to Tecnopolis to learn about a successful program of this type in Waterloo, Canada.

In the three days of the present workshop, I hope to obtain the answer to some of these questions. Some successful and interesting programs in Europe and in the U.S. will be presented. I expect this international workshop to be a model for similar efforts in the future. My congratulations to SASIAM and TECNOPOLIS for organizing this exciting workshop.

14. Mathematical Simulation and Optimization of Chemical Plants

F. Kokert, L. Peer, Hj. Wacker

14.1 Introduction

In recent years considerable effort has been devoted to the mathematical simulation of chemical plants using flowsheeting techniques. There are three stages to consider:

i) synthesizing a total flowsheet (see e.g. M.D.Lu, R.L.Motard [8]

ii) Mathematical Simulation (see e.g. H.P.Hutchinson, D.J.Jackson, W.Morton [6], L.M.Timàr, F.Simon, Z.Csermely, J.Silklós, S. Bacskai, J.Edes [12] and Fredenslund [5])

iii) Optimization

In this paper we only deal with ii) and iii) and in addition we consider only steady state processes. The aim of our work is to study and improve the mathematical algorithms especially with respect to numerical aspects. To sketch the idea what we want to discuss we present the highly simplified flowsheet of an Oxygen Plant.

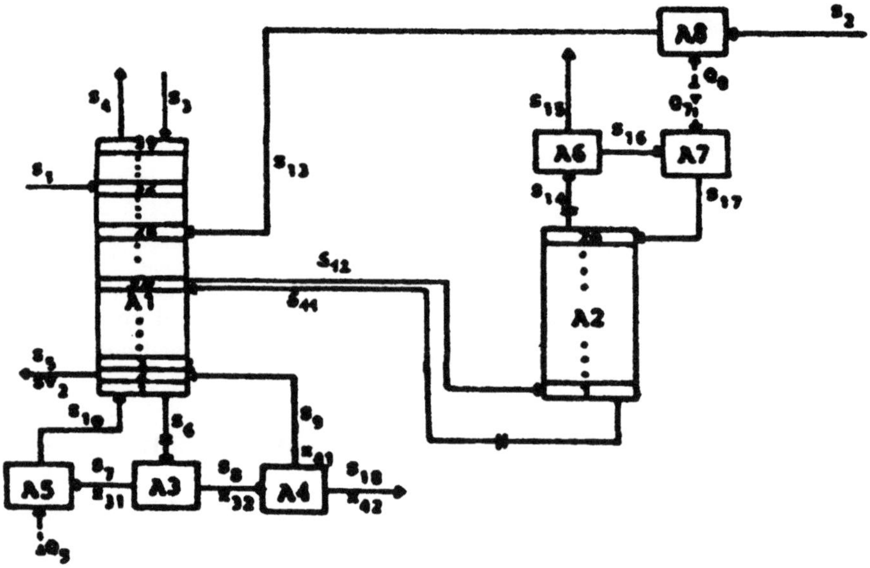

Figure 55. Oxygen Plant

The flowsheet consists of a number of units: plate absorption columns (A1, A2), heater (A5, A6), cooler (A7), splitter (A3, A4, A6) and connecting streams (18). Our model is based on mass balances, energy balances, liquid/vapor equilibrium conditions.

The specific form of the functions depends on the underlying thermodynamic model, e.g. UNIQUAC.

14.2 Mathematical Simulation of Chemical Plants - Flowsheeting

Even the simplified example above gives an idea why the Mathematics behind flowsheeting problems might be cumbersome. Beside simple units like mixers, splitters, etc. a flowsheet may contain heat exchangers, reactors and plate distillation columns. The mathematical model for, say, a plate distillation column with N plates and M components leads to a nonlinear equational system with $(2M + 1) \cdot N$

variables. Even for some few components some hundred variables have to be solved.

14.2.1 Flowsheeting

The flowsheet of a chemical plant can be seen as a directed graph where the knots are defined by the units and the edges are given by the (directed) streams. The graph is described by the incidence matrix $I(G)$:

$$I(G)_{ij} := \begin{cases} 1 \text{ if there is an edge from knot i to knot j;} \\ 0 \text{ else} \end{cases}$$

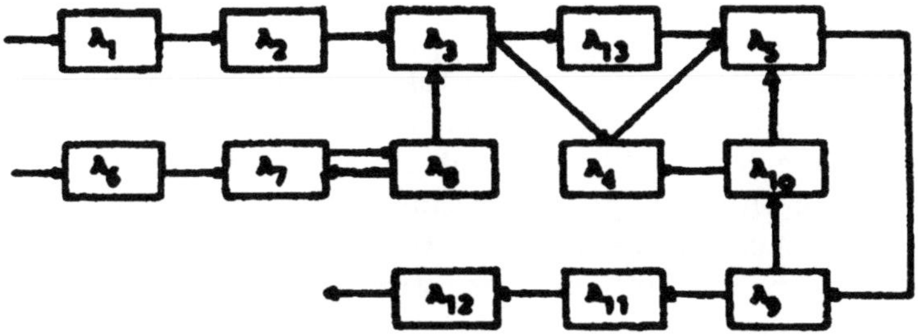

Figure 56. Example of a Flowsheet

A chemical plant, resp. a flowsheet, can be determined sequentially if there exists a numbering of the units such that the resulting incidence matrix becomes right upper triangular. In general, however, most chemical plants involve feed back connections. Therefore, only certain subsets of the flowsheet can be determined sequentially. Technically this means that the incidence matrix is ordered in such a way that there results a right upper block triangular matrix. It is required that the diagonal blocks cannot be split up into smaller subblocks. This means that we have to search for irreducible subsystems (IRS) of the graph. To do this we need elementary cycles. An elementary cycle is a path where the starting knot coincides with the ending knot. All knots of the cycle occur only once except the first and the last one. For our example we have the following three elementary cycles: $\{A_7, A_8\}, \{A_5, A_9, A_{10}\}, \{A_5, A_9, A_{10}, A_4\}$.

Tiernan [11] gave an algorithm by which one can find all elementary circuits of a directed graph. It can be shown that an IRS is either a set consisting of one element only or it is a union of elementary cycles. Irreducible here means that the subsystem is not allowed to split up into jointly void union of elementary circuits. In our example we have the following IRS: $\{A_1\}$, $\{A_2\}$, $\{A_3\}$, $\{A_4\}$, $\{A_7, A_8\}$, $\{A_4, A_5, A_9, A_{10}\}$, $\{A_{11}\}$, $\{A_{12}\}$.

When including restrictions (e.g. a certain temperature is specified etc.) it might be necessary to unite certain subsystems in order to make the number of free parameters correspond to the number of restrictions. In most practical cases the resulting IRS's are still large because of feed back structures. There are two main strategies to cope with this situation.

a) *Equational Approach*
Each IRS is described and solved in total by one (i.g. large) nonlinear equational system. Numerical experiments clearly show that this approach is highly efficient. As a draw back one finds that available library programs for units (reactors, heat exchangers etc.) cannot be used directly but must be included in the system. An example of this strategy is given by the QUASILIN package proposed by Hutchinson et al [6]

b) *Sequential Modular Approach*
The IRS is split up into smaller parts by cutting certain connecting streams. Starting values are given to these torn streams such that the IRS can be determined sequentially. As an example we consider the IRS $\{A_4, A_5, A_9, A_{10}\}$ from Figure 56.

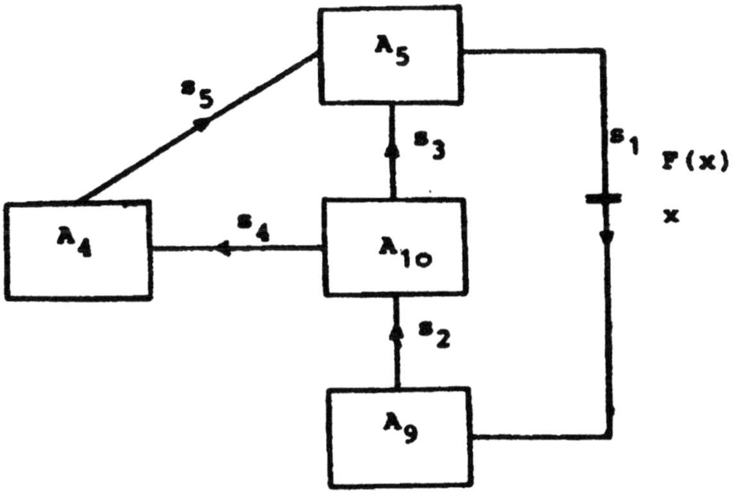

Figure 57. An Irreducible Subsystem

The stream s_1 (from A_5 to A_9) is split up. Starting with x as input. The IRS can then be computed sequentially: $x \rightarrow A_9(x) \rightarrow A_{10}^1(A_9(x)) \rightarrow A_4(A_{10}^1(A_9(x)))$, $A_5 = A_5(A_4, A_{10}^2)$. To preserve continuity we have to observe the fixed point equation $x = F(x)$. An IRS therefore is solved by two types of iterations:

i) the outer cycle is determined by the cutting streams leading to a fixed point type nonlinear equation;

ii) the inner cycle is defined by the units which themselves are usually solved iteratively.

The spitting can be done automatically, see e.g. Pho/Lapidus [9].

Of course it is up to one's own decision what is defined as a unit. In our program, e.g., both plate distillation columns with and without extensions like side strippers and/or pumparounds can be defined as units, for instance.

14.2.2 Newton's Method for the Solution of Irreducible Subsystems

In the following we want to propose a special solution technique based on the sequential modular approach. We want to solve the system

$$x = F_k(x) \quad k = 1, \dots, n \quad (n\text{: number of nontrivial IRSs})$$

In the case of restrictions there are additional parameters $P = (P_1, \dots, P_r)$ to be determined. Examples for restrictions are given purity of the top product of a plate distillation column at unknown reflux or given output temperature of a mixer at an unknown heat input etc. One gets:

$$(1) \qquad \begin{cases} x - F(x) = 0 \\ R(x,P) = 0 \end{cases}$$

We assume that we have at least directional derivatives both for the units and the restriction function R. We have to make sure that we have a 'legal' set of variables/equations, i.e. it is necessary that the 'input' variables together with the specified variables uniquely define the 'output' variables. Under our assumption we may use Newton's method. The main effort when using Newton's method is to be devoted to get the Jacobian. To do this analytically is out of reach: to determine a heat exchanger, for instance, a nonlinear twopoint boundary value problem has to be solved which can be done only numerically. Because of the numerical effort also divided differences are to costly. Under the assumption that the gradients of each single unit are available - this can be organized when computing the units - the Jacobian is easily determined by the help of the chain rule. We demonstrate this for the following IRS:

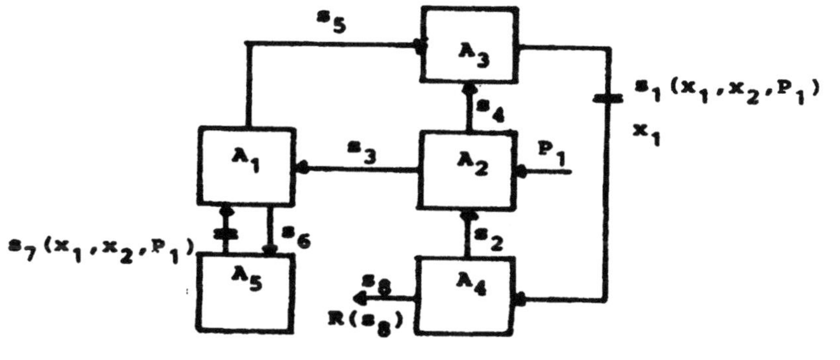

Figure 58.

The system consists out of 5 units. In unit A_2 we have the free parameter P_1 which corresponds to the restriction $R(s_8)$ in dependence on the output stream s_8 . We have three elementary cycles : $\{A_1, A_5\}$, $\{A_1, A_3, A_4, A_2\}$, $\{A_3, A_4, A_2\}$. Using the tear streams $\{s_1, s_7\}$ we get:

$$\text{(2)} \qquad \begin{aligned} s_1(x_1, x_2, P_1) - x_1 &= 0 \\ s_7(x_1, x_2, P_1) - x_2 &= 0 \\ R(x_1, x_2, P_1) \qquad &= 0 \end{aligned}$$

To get the influence of x_i on s_j : $\partial s_j / \partial x_i$ we consider all paths starting at x_i and ending at the cutting stream s_j. The sensitivity factor $\partial s_j / \partial x_i$ is then given by summing up the products of all sensitivity factors of each path. To get $\partial s_1 / \partial x_1$ for instance, we have the two paths: $x_1 \to s_2 \to s_4 \to s_1$ and $x_1 \to s_3 \to s_5 \to s_1$:

$$\frac{\partial s_1}{\partial x_1} = \frac{\partial s_1}{\partial s_4}\frac{\partial s_4}{\partial x_1} + \frac{\partial s_1}{\partial s_5}\frac{\partial s_5}{\partial x_1}$$

further

$$\frac{\partial s_4}{\partial x_1} = \frac{\partial s_4}{\partial s_2}\frac{\partial s_2}{\partial x_1} \quad \text{and} \quad \frac{\partial s_5}{\partial x_1} = \frac{\partial s_5}{\partial s_3}\frac{\partial s_3}{\partial s_2}\frac{\partial s_2}{\partial x_1}$$

In the case where restrictions are involved one has to use (2), for instance:

$$\frac{\partial R}{\partial x_1} = \frac{\partial R}{\partial s_8}\frac{\partial s_8}{\partial x_1} \quad \text{and} \quad \frac{\partial R}{\partial x_2} = 0$$

The numerical effort compared to the evaluation of the flowsheet (= function evaluation) is small. Newton's method then gives a correction $\Delta z_k = (\Delta x_k, \Delta P_k)$. We use a damped version:

$$z_{k+1} = z_k + \lambda_k \Delta z_k$$

In addition it might be that some variables must be projected to remain feasible, i.e. $T(^{\circ}K) \geq 0$ etc.

14.2.3 Gradients of units

Some simple units like mixers, splitters etc. allow for an analytical computation of gradients. For more complex units like plate distillation columns, flash drums, compressors, heat exchangers, reactors the chain rule again proves useful.

14.2.3.1 Numerical Determination of Gradients

For demonstration we discuss a flash unit

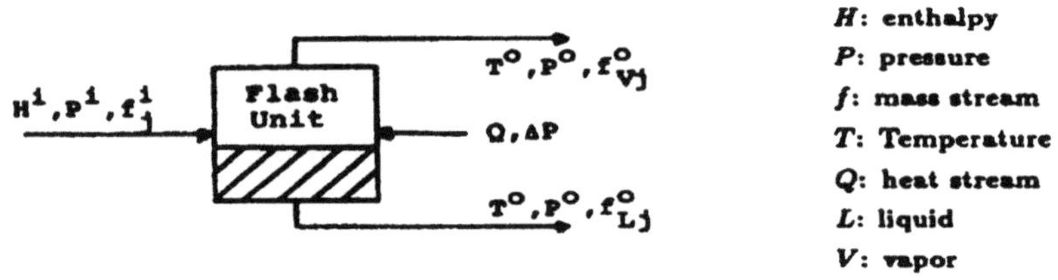

Figure 59. Flash Unit

A given input stream (H^i, P^i, f^i_j) is split into vapour (V) and liquid (L) components. The thermodynamic equilibrium is controlled by the heat flux Q and the variation of pressure ΔP. We get the following system:

$$P^o = P^i - \Delta P \qquad \begin{array}{l} \text{i : input} \\ \text{o : output} \end{array}$$

i) mass balances:
$f^o_{Lj} + f^o_{Vj} - f^i_j = 0, \quad j = 1, \dots, N$

ii) energy balance:
$H_L(f^o_{L1}, \dots, f^o_{LN}, T^o, P^o) + H_V(f^o_{V1}, \dots, f^o_{VN}, T^o, P^o) - H^i - Q = 0$

iii) equilibrium conditions:

$$k_j(f^o_{L1}, \dots, f^o_{Ln}, f^o_{V1}, \dots, f^o_{VN}, T^o, P^o)\cdot\frac{f^o_{Lj}}{L} - \frac{f^o_{Vj}}{V} = 0, \quad j = 1, \dots, N$$

where

$$L := \sum_{i=1}^{N} f^o_{Li}, \quad V := \sum_{i=1}^{N} f^o_{Vi}$$

(i) - (iii) give a nonlinear system $F(x,p) = 0$ with the variables $x = (f^o_{L1}, \dots, f^o_{VN}, T)$ and the parameter $p = (H^i, P^i, f^i_1, \dots, f^i_n, Q, \Delta P)$. We use the identity $F(x(p),p) = 0$ where $x(p)$ solves our system in dependance on p. We have:

$$(\bullet) \qquad 0 = \frac{dF(x(p),p)}{dp} = \frac{\partial F(x(p),p)}{\partial x}\cdot\frac{\partial x(p)}{\partial p} + \frac{\partial F}{\partial p}$$

As we use Newton's method to solve $F(x,p) = 0$ for x the Jacobian $\partial F / \partial x$ is already (even factorized) at our disposal. Therefore, one only has to compute $\partial F / \partial p$ and solve $(\bullet)$ for $\partial x(p) / \partial p$.

In the special case of a flash unit $\partial F / \partial H^i$, $\partial F / \partial dQ \ (= \partial F / \partial H^i)$, $\partial F / \partial f^i_j$ are determined analyticaly whereas $\partial F / \partial P^i \ (= \partial F / \partial P^o)$, and $\partial F / \partial \Delta P \ (= -\partial F / \partial P^i)$ may be approximated by forward differences.

14.2.4 Special Units

In the following we give some examples of units which are of a more complex nature.

14.2.4.1 Plate Distillation Columns

We consider configurations having one main column with a few pumparounds and a small number of side strippers which are linked with the main column (see Figure 60). The mathematical model consists of three sets of equations.

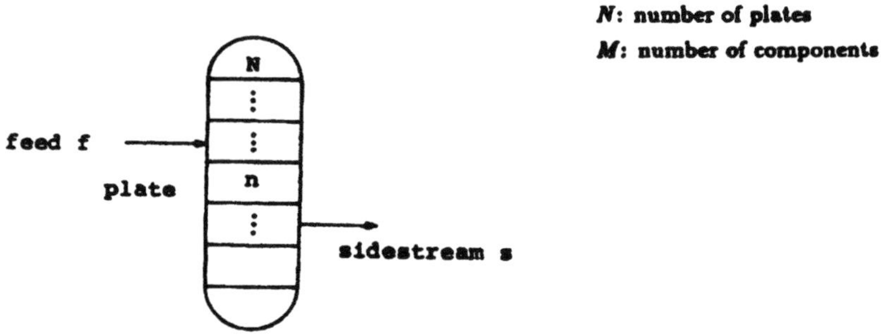

Figure 60. Main Column

Component mass balances (all plates, all components)

$$b_{nj} \equiv \left(1 + \frac{S_n^L}{L_n}\right) l_{n,\,j} + \left(1 + \frac{S_n^V}{V_n}\right) v_{n,\,j} - l_{n+1,\,j} - v_{n-1,\,j} - f_{n,\,j} = 0$$

$$L_n = \sum_j l_{n,\,j} \qquad V_n = \sum_j v_{n,\,j}$$

Enthalpy balances: (all plates)

$$E_n \equiv (L_n + S_n^L)H_n^L + (V_n + S_n^V)H_n^V - L_{n+1}H_{n+1}^L - V_{n-1}H_{n-1}^V - F_nH_n^L - \Delta E_n = 0$$

Vapor-liquid equilibrium relationship: (all plates, all components)

$$g_{n,\,j} \equiv K_{nj}\frac{l_{nj}}{L_n} - \frac{v_{nj}}{V_n} = 0$$

In addition H_n^L, H_n^V and K_{nj} are given functions of (known) pressure, temperature and the composition of the liquid and vapor phase. Their particular form is based on the UNIQUAC model.

$$H_n^L = \sum_j (a_j + b_j T) \frac{l_{nj}}{L_n} + H_n^E(l_{n,1}, \dots, l_{n,m}, T_n)$$

$$H_n^V = \sum_j (c_j + d_j T_n) \frac{v_{nj}}{V_n}$$

$$k_{nj} = \frac{par_j(T_n)}{P_n} \cdot \frac{\gamma_j(l_{n1}, \dots, l_{nM}, T_n)}{\phi_j(v_{n,1}, \dots, v_{n,M}, T_n)}$$

It is easy to see that by ordering the unknowns resp. equations of plate n in the way $(l_{n,1}, \dots, l_{n,M}, T_n, v_{n,1}, \dots, v_{n,M})$ resp. $(b_{n1}, \dots, b_{nM}, E_n, g_{n1}, \dots, g_{nM}) = 0$ and by arranging these blocks according to the natural ordering of the plates the Jacobian results to be N-by-N block tridiagonal with $(2M + 1)$-by-$(2M + 1)$ submatrices.

In the case of extension like pumparounds and/or side strippers there arise additional blocks outside the block tridiagonal part of the Jacobian. The numerical solution is described in [1]. An analysis of the numerical effort and numerical experiments show that the system should be linearized by Newton's method (no Quasi Newton technique!). The linearized system can be solved by L-U factorization with incomplete partial pivoting under exploitation of the special structure of the Jacobian. As an alternative one may use iterative methods which proved superior especially in the case of extensions. Both these methods were favorably compared with existing solution techniques, see [1], [13].

14.2.4.2 Heat Exchanger

A counterflow heat exchanger consists of a system of N parallel tubes through which liquid and/or vapor mixtures of different temperatures flow countercurrently. The input temperature T_i^* of each stream is known, the output temperature is to be determined. Figure 61 gives the principle of this type of heat exchanger.

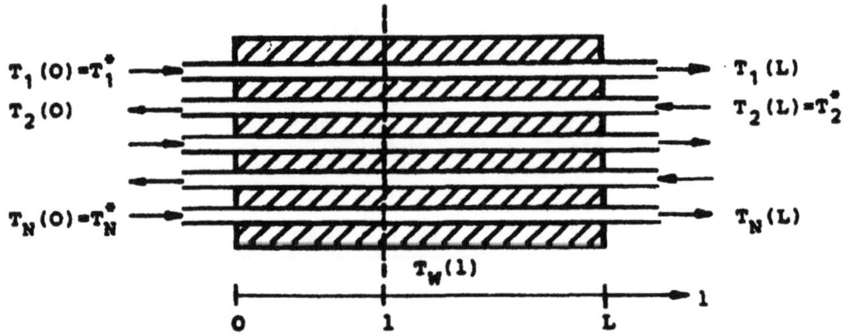

Figure 61. Heat Exchanger

We assume

i) there is no heat transport in length direction inside the tubes except the heat transport by mass flow;

ii) for l given we have one joint temperature T_w of the apparatus.

The steady state is characterized by balancing the heat fluxes for each mixture

$$\frac{d}{dl}[m_i H_i(m_{ij}, T_i)] = \Delta Q_i = k_i F_i(T_w(l) - T_i(l))$$

An overall heat balance gives:

$$\sum_i sign(R_i)\Delta Q_i + k_w F_w [T_w(l) - T_A] = 0$$

There results the following nonlinear BVP with separated boundary values:

$$\frac{dT_i}{dl} = \frac{\dfrac{\sum_{j=1}^{N} T_j(l)\, F_j(l)\ sign(R_j) + T_A k_w F_w}{\sum_{j=1}^{N} k_j F_j + k_w F_w} - T_i(l) - m_i \dfrac{\partial H_i}{\partial p_i} \cdot \dfrac{\partial p_i}{\partial l}}{m_i \cdot \dfrac{\partial H_i}{\partial T_i}}$$

$T_i(0) = T_i^*$ $i \in J$

$T_i(L) = T_i^*$ $i \in \{1, \dots, N\}/J$, T_i, T_A : temperature of tube i resp. working hall

F_j, F_w : periphery of tube i resp. wall

$$sign(R_j) = \begin{cases} 1 & \text{stream } j \text{ goes in "main" direction} \\ -1 & \text{stream } j \text{ goes against "main" direction} \end{cases}$$

k_i, k_w : heat conduction coefficient of tube i resp. wall

H_i : specific enthalpy of stream i

p_i : pressure of stream i

m_{ij} : composition of stream i

In the case of phase transitions within the heat exchanger the right hand side of the ODEs may have a jump at some a priori unknown point l. Even the number of jumps (at most two for each flow) may change during iteration. We have to adapt the classical multiple shooting technique (MST) to this situation:

a) Starting procedure: for each stream we determine both the dew and the bubble point curves $T^d(l)$ and $T^b(l)$ resp.

b) MST iteration: assume that we have a phase transition for stream i during an iteration, e.g. $T_i(l_k) < T_d(l_k)$ and $T_i(l_{k+1}) > T_d(l_{k+1})$. Though we want to use the fixed grid which defines MST we are free to use the unknown dew point $l_k + \delta$ as an integration point. δ is determined by adding the nonlinear equation $T_i(l_k + \delta) = T_i^d(l_k + \delta)$ to the nonlinear system resulting from the BVP.

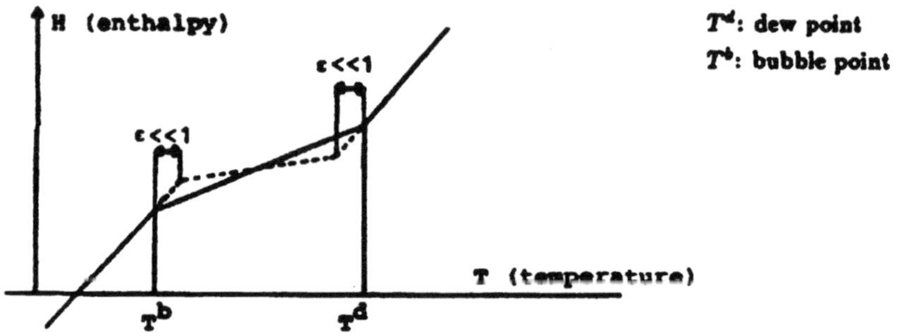

Figure 62.

To avoid convergence problems the two phase region is artificially reduced (see Figure 62). In case of a two phase situation during the iterations a flash-drum computation has to be performed. At given temperature and pressure we use mass balances and the equilibrium conditions. The condition of the multiple shooting matrix is extremely ill conditioned. Therefore, the classical recursive procedure fails. Instead we used total column pivoting under observation of the special structure of the Jacobian. The numerical amount increases considerably (up to six times compared with classical MST proposed by Bulirsch/Stoer). For details see Leitner [7].

Example: We consider a heat exchanger with 3 countercurrent streams of (different) Argon-Oxygen-Nitrogen mixtures. We confine ourselves to give results mainly on phase transitions:

stream 1 ($\leftarrow$), $T_1^* = 190°K$

stream 2 ($\rightarrow$), $T_2^* = 95°K$ sum of enthalpies of inlet streams: -2035675 [J/h]

stream 3 ($\rightarrow$), $T_3^* = 105°K$

1. Iteration: no phase transition; residuum $r = 0.26 \cdot 10^2$

2. Iteration: phase transition (T^b) for stream 2 at $l = 0.273$; $r = 0.28 \cdot 10^1$

3. Iteration: phase transition (T^b) for stream 2 at $l = 0.277$; $r = 0.51 \cdot 10^1$

.

.

7. Iteration: phase transition (T^b) for stream 2 at $l = 0.289$; $r = 0.19 \cdot 10^{-4}$

Results: (output temperature)

stream 1: 128.94°K

stream 2: 98.54°K sum of enthalpies of outlet streams: -1 990 312 [J/h]

stream 3: 121.61°K

For stream 2 we give also the distribution of the components: [mol/h]

l=0	Ar	N_2	O_2
liquid	10	71	20
volatile	-	-	-

l=1	Ar	N_2	O_2
liquid	7.93	46.37	16.41
volatile	2.07	24.63	3.59

Remark : To get the gradients one needs, roughly, 10 % of the total numerical amount.

14.2.4.3 Tubular Reactors

We discuss reactions of the following type:

$$\sum_{i=1}^{N} a_i^R A_i \rightarrow \sum_{i=1}^{N} a_i^P A_i$$

a_i^P, a_i^R: stoichiometric constants
A_i: reaction components (N components)

We assume that the concentration of the reaction components velocity is proportional to the concentration of the reaction components (time law). Further we confine ourselves to simple reactions, i.e. we assume that the stoichiometric constants give directly the exponentials in the time law.

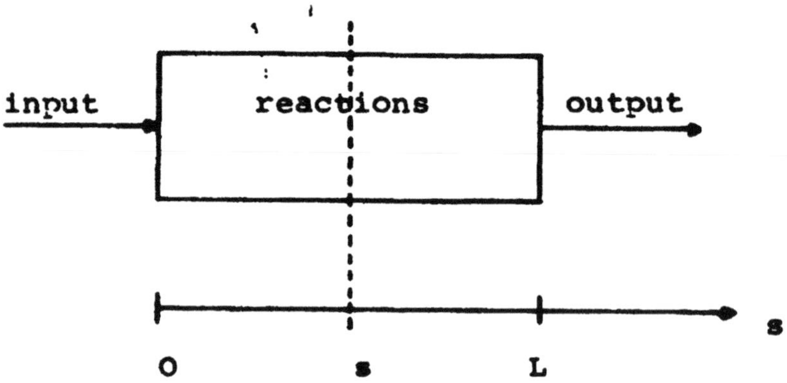

Figure 63. Tubular Reactor

Though we only are interested in the stationary case we describe the process dynamically, i.e., in dependence on the position s in the reactor. We get:

$$\frac{d\xi_j}{ds} = A\cdot(k_j^R(T)\prod_{i=1}^{N} c_i^{a_{ij}^P} - k_j^P(T)\prod_{i=1}^{N} c_i^{a_{ij}^R}) \quad j = 1, \dots, M$$

$$\frac{dT}{ds} = \frac{-\sum_{i=1}^{N}\frac{dH_g}{dm_{ig}}\cdot\frac{dm_i}{ds} + \sum_{i=1}^{N}\frac{dH_s}{dm_{is}}\cdot\frac{dm_{is}}{ds} + \frac{dH_F}{dp}\cdot\frac{dp}{ds} + k_w\cdot U(T - T_w)}{\frac{dH_F}{dT}}$$

$$\xi_i(0) = 0 \qquad T(0) = T_0$$

From this we determine the mass streams m_i

ξ: reaction intensity (M reactions);

A: cross section of the tube;

c: concentration;

U: periphery of the tube;

P,R: direction of reaction

k: velocity constant

H_g, H_s, H_F: Enthalpy of gases, resp. solids, resp. components

k_w, T_A: conductivity of the wall, outside temperature

This system of ODEs may be extremely stiff. For the numerical solution we use the semi implicit midpoint rule proposed by Deuflhard [3]. We confine ourselves to give an example, for details see [2].

$$CO + H_2O \leftrightarrow CO_2 + H_2$$

$$2COS \leftrightarrow CO_2 + CS_2$$

$$\frac{d\xi_1}{ds} = A\cdot(k_1^P c_{CO_2} c_{H_2} - k_1^R c_{CO} c_{H_2O})$$

$$\frac{d\xi_2}{ds} = A\cdot(k_2^P c_{CO_2} c_{S_2} - k_2^R c_{cos}^2)$$

[Kmol/h]	CO	CO_2	H_2O	COS	H_2	CS_2
Input	4.15	4.15	2.8	0.5	-	-
output	1.35	6.95	0.00	0.05	2.80	0.00

s	ξ_1	ξ_2	T
$0.5\cdot10^{-6}$	0.01297	$0.5\cdot10^{-7}$	1316
$0.4\cdot10^{-4}$	0.81981	$0.4\cdot10^{-5}$	1367
$0.4\cdot10^{-3}$	2.47336	$0.4\cdot10^{-5}$	1465
$0.3\cdot10^{-2}$	2.79914	$0.3\cdot10^{-5}$	1483
1.	2.79930	$0.3\cdot10^{-5}$	1483

The reaction is finished already at s = 0.0005 m. The number of integration steps (13) corresponds to the number of evaluations of the Jacobian. Computing time: 15.8 sec (BASF 7/78).

14.2.5 Example: Oxygen Plant

An oxygen plant splits up air into its components: nitrogen, oxygen and argon. The flowsheet is given in Figure 64

Units: plate absorption column (AC) (3), mixer (M), cooler (C), heater (H) (17), splitter (S) (10), compressor (Co) (2); in addition dew / bubble point computations are simulated by bubdew (BD) (5) (dotted).

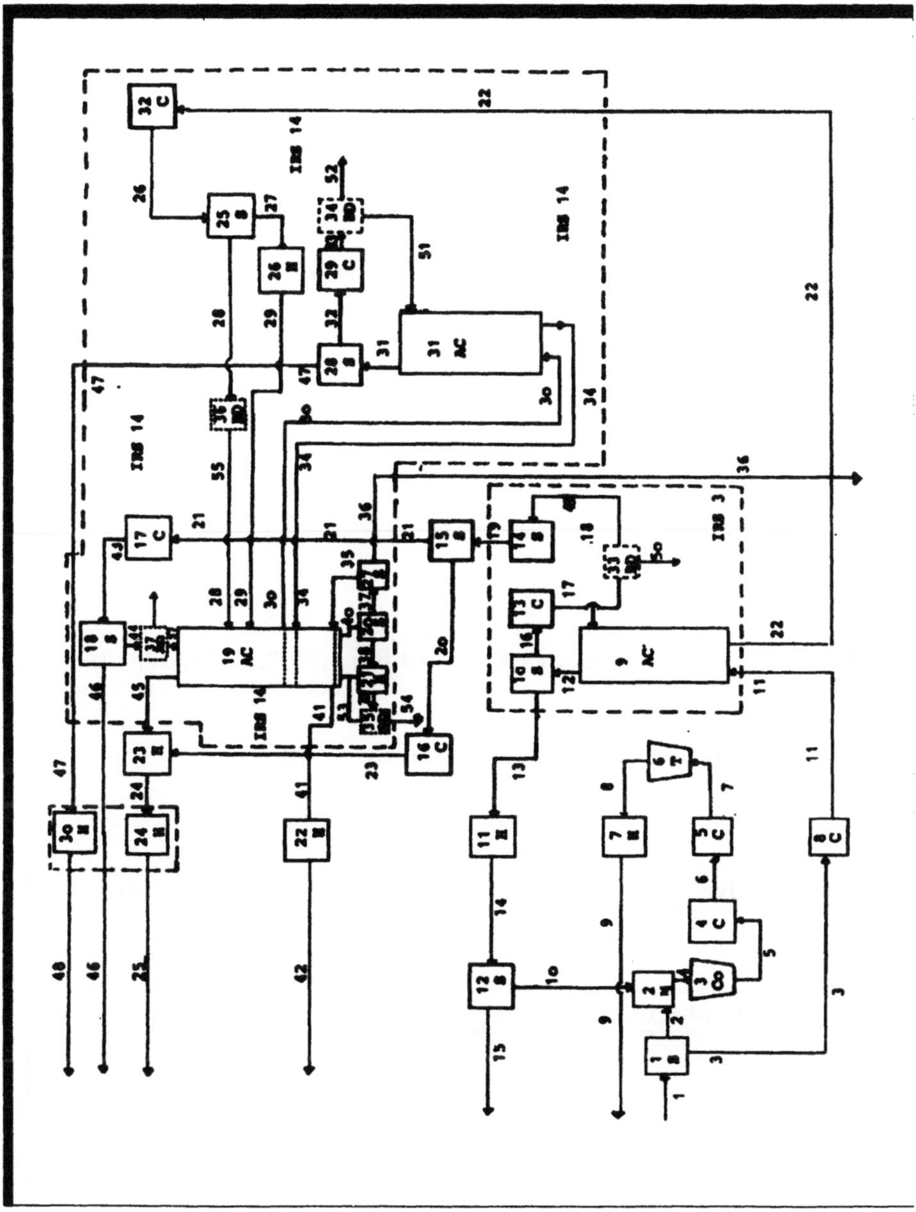

Figure 64.

Streams: feed (1), product streams (12), connecting streams (45)

Restrictions: 25

e.g.

$T_{11} - 108 = 0$ (temperature)

$Q_5 + Q_7 + Q_8 + Q_{11} + Q_{22} + Q_{24} + Q_{30} = 0$ (heat)

$D_{29} - 0.9999 = 0$ (stream 29 is mostly vapor)

$F_{13} - 655.4 = 0$ (total flowrate of stream 13 is given)

$x_{12}^{16} + x_{12}^{13} = 1$ (stream 12 is split up into stream 16 and stream 13)

.

.

Free Parameters: 25

a) Heating/Cooling duties at the units: 4, 5, 7, 8, 11, 13, 17, 21, 22, 23, 24, 26, 29, 30, 32

b) Distribution factors at the units: 10, 14, 20, 25, 28

The flowsheet can be split up into 17 IRSs:

IRS	Units	Tear Streams	Restrictions
1	1	-	-
2	8	-	1
3	33,14,9,10,13	17	17,18,21,22,9
		. . .	
16	23	-	16
17	24,30	-	7,8

In the flowsheet the nontrivial IRSs 3 and 14 are given by dotted lines.
We confine ourselves to present some few results: IRS 14:

Tear Stream: 30(Ar, N_2, O_2) :

starting point 12.81, 16.22, 257.80 [mol/h]

result (6 It.) 14.85, 0.12, 290.83 [mol/h]

Product stream 42:

Ar: 37.08, N_2:0.09, O_2:945.32 [mol/h]

Enthalpy: $-0.18544.10^6$ [J/h] , Temp.: 292 [°K];

Computing time: 1′14″ (BASF 7/78)

14.3 Optimization

There are two aspects of optimization with respect to flowsheeting

i) One might be interested in the structure of a flowsheet (e.g. inserting a heat exchanger, determining the number of plates for a distillation column etc.) and

ii) one might like to know how to choose certain process parameters (e.g. reflux of a distillation column, input of heat etc.) such that certain goals (e.g. purity of products, costs for heating etc.) are achieved as good as possible. Here we only deal with the second aspect.

In Chapter 14.2, "Mathematical Simulation of Chemical Plants - Flowsheeting" we described how a function evaluation for the steady state of a chemical plant is to be performed by Newton's method. The Jacobian is determined effectively by the chain rule. If we want to make full use of this solution technique we use Sequential Quadratic Optimization techniques, compare f.e. Fletcher [4]. $G^{(k)}$ is approximated by Quasi Newton technique making full use of the gradients.

$$f(x+s)^{(k)} \approx q^{(k)}(s) := f^{(k)} + s^T g^{(k)} + \frac{1}{2} s^T G^{(k)} s$$

$$q^{(k)}(s) = \text{Min!}$$

$$a_i^T s = b_i - a_i^T x^{(k)} \quad i \in E \quad \text{(equality constraints)}$$

$$a_i^T s \geq b_i - a_i^T x^{(k)} \quad i \in J \quad \text{(inequality constraints)}$$

In our case we only consider box-constraints for the parameters. To achieve convergence we use trust region technique.

We confine ourselves to give an example, for details see [10].

Example: Optimization of a methanol-ethanol-water plate distillation column.

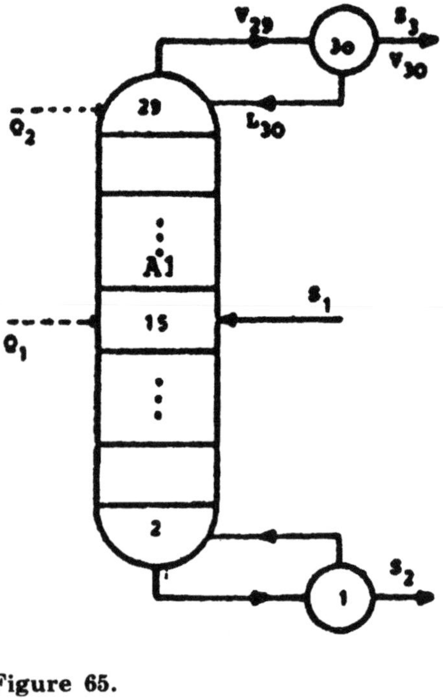

Figure 65.

Objective: We want to minimize the percentage of water in the distillate:

$$f(x) \equiv x_{H_2O}(DEST, RFLX, Q_1, Q_2) = \text{Min!}$$

distillate: $DEST := V_{30}$

Reflux relation: $RFLX := L_{30}/V_{30}$
heat influx: Q_1 (plate 15), Q_2 (plate 29)

Constraints:

$$1 \leq Dest < 20 \quad [mol/s] \qquad 0.2 \leq RFLX \leq 30$$

$$-1 \leq Q \leq 1 \qquad [J/s] \qquad 0 \leq Q_2 \leq 300000 \quad [J/s]$$

Result:

$$x_0 = (19,10,0,0), \quad f_0 = 7.16969$$

$$x_* = (8.4215,\ 30,\ 1.84 \cdot 10^{-4},\ 0)^T, \quad f_* = 0.00211$$

Acknowledgement : Research was supported by the Austrian Science Foundation Fonds (projects P5729 and P6674P), the Austrian Ministry of Research (sect. Research) and the VOEST ALPINE AG (Abt. FAT).

14.4 References:

[1] D. Auzinger, L. Peer, Hj. Wacker, W. Zulehner: Numerical Calculation of Separation Process, in: Case Studies in Industrial Mathematics, ECMI II, (ed. Engl / Wacker / Zulehner), 1988, pp. 131-154

[2] G. Bachleitner: Simulation of a Tubular Reactor, Diploma Thesis, Univ. Linz, Math. Dept., Febr. 1988, pp. 1-105 (in cooperation with VOEST - ALPINE AG, Abt. FAT) (in German)

[3] P. Deuflhard: Recent Progress in Extrapolation Methods for ODEs., Univ. Heidelberg, Techn. Report Nr. 224 (1983)

[4] Fletcher: Practical Methods of Optimization, Vol. 2: Constrained Optimization, John Wiley & Sons, New York, pp. 1-224

[5] A Fredenslund, J. Gmehling, P. Rasmussen: Vapor-Liquid Equilibria using UNIFAC, Elsevier, Amsterdam, 1977

[6] H. P. Hutchinson, D. J. Jackson, W. Morton: The Development of an Equational Oriented Flowsheet Simulation and Optimization Package I, II, Comp. Chem.Eng. Vol. 10, Nr. 1, (1986), pp. 19-29, 31-47

[7] M. Leitner: Mathematical Simulation of a Matrix Heat Exchanger, Diploma Thesis, Univ. Linz, Math. Dept., August 1988 (in German)

[8] M D. Lu, R. L. Motard: Computer-Aided Total Flowsheet Synthesis. Comp. Chem. Eng., Vol. 9, Nr. 5, (1985), pp. 431-445

[9] T. K. Pho, L. Lapidus: An Optimum Tearing Algorithm for Recycle Systems, AICHE Journal, vol. 19, Nr. 6 (1973), pp. 722-726

[10] J. Synka: Optimization of Chemical Plants with Box Constraints (Plate Distillation Columns), Diploma Thesis, Univ. Linz, Math. Dept., June 1988, pp. 1-150 (in German)

[11] J. C. Tiernan: An efficient search algorithm to find elementary cycles of a Graph, Comm. of the ACM, Vol. 13, Nr. 12, 1970, pp. 722-726

[12] L. Timàr, F. Simon, Z. Csermely, J. Siklós, S. Bàcskai, J. Edes, Comp. Chem. Eng., Vol. 8, Nr. 3/4, (1984), pp. 185-194

[13] Hj. Wacker: Mathematical Research and Industrial Consulting in Linz: Power Plants - Chemical Columns - Heating of Slabs, in: 1st Europ. Symp. on Math. in Industry, ECMI I (eds. Hazewinkel/Mattheij/van Grossen), 1988, pp. 163-185

15. The Use of Mathematical Methods in Production Management

W.H.M. Zijm[4]

15.1 Abstract

In this report, the use of mathematical models and mathematical techniques for solving design and planning problems in industrial production systems is discussed. We describe several projects carried out in different Philips factories. Topics include design problems in an (automated) manufacturing system, production planning and inventory management in a Telecommunication company, and shopfloor scheduling problems in a cable factory. In addition, we briefly list a number of research activities, motivated by these projects, which take place at the Eindhoven University of Technology.

15.2 Introduction

The development of new industrial products and of advanced technologies requires an ever increasing application of, often complex, mathematical techniques. As an example, one may think of the use of combinatorial methods for the design of integrated circuits (IC's), the use of discrete mathematics for the development of coding systems, the finite element method as a building stone in certain CAD (Computer Aided Design) systems, applications of systems theory in audio and video signal processing, developments in fluid

[4] Nederlandse Philips Bedrijven B.V.,
Centre for Quantitative Methods, Building HCM-721,
p.o. Box 218, 5600 MD - Eindhoven, The Netherlands,

and

Eindhoven University of Technology,
Department of Mathematics and Computing Science,
Building DG-011, p.o. Box 513, 5600 MB - Eindhoven,

dynamics, etc. To quote David(1984): "When we entered the era of high technology, we entered the era of mathematical technology."

Apart from their use in approaching purely technical problems, there is a growing tendency to apply mathematical techniques for designing, planning and controlling complex industrial processes, with the aim to increase their performance. In particular, Industrial Statistics and Operational Research have provided useful tools that can be applied in this area. Mathematical statistics, arosen initially from an attempt to describe certain processes in demography (Malthus) and biogenetics (Pearson, Fisher) became popular in industry by the work of Walter Shewhart on statistical quality control at Bell Laboratories. Other pioneers in this field are e.g. Wald, Deming, Juran and Crosby.

The term "Operations Research" or "Operational Research" stems from the second world war, when scientific methods were developed to solve complex logistics problems. However, industrial mathematics that can be classified as Operational Research avant la lettre, can be traced back to the preceding century. In 1832, Charles Babbage, who later became famous as the father of the first digital computer, wrote "On the Economy of Machinery and Manufactures" in which he followed and extended Adam Smith's ideas on labour division. Another pioneer was the Danish mathematician Erlang; in his studies on the expected performance of telephone exchanges he developed the roots of modern queueing theory. Stochastic networks were proposed by Jackson in the early sixties to study the behavior of Job Shop production systems (cf. Jackson(1963)). In the field of production control, the book of Holt, Modigliani, Muth and Simon "Planning Production, Inventories and Work Force" marked an important step ahead (see Holt et. al.(1960)). Forrester's "Industrial Dynamics" is now recognized as a pathbreaking study on the cyclical variation of stocks in large production/distribution chains (cf. Forrester(1961)).

Despite of all this, the acceptance of mathematics as an important tool to solve complex industrial problems is certainly not as widespread as seems to be desirable. Industrial mathematicians are working mainly within large companies, and quite often in a research function, developing mathematical methods for the design of advanced products and complex technologies. The use of mathematics on a routine basis as a tool for planning and controlling complex production processes is still limited, despite the undeniable successes that have been reached. Only large multinational organisations (e.g. IBM, AT&T, Philips) seem to employ groups of mathematicians who direct their efforts primarily to the solution of problems arising in the field of production management and logistics control.

Philips' Centre for Quantitative Methods is a group of mainly mathematically skilled consultants working on projects in the latter field: industrial production management, industrial process control (including quality control), logistics issues, design and control of flexible manufacturing systems, forecasting and project management, etc. In this report, we describe some projects carried out in different Philips factories by the author, as a member of the Operations Research Group of the Centre for Quantitative Methods. The examples each highlight a specific type of problem. Production planning and inventory management is the key element in the analysis of the Production of telephone exchanges in a Telecommunication company. Certain design problems had to be solved when developing a completely automated production line for transformers (to be built in in TV sets). Shopfloor scheduling techniques were proposed for a mains leads department in one of Philips' cable factories.

In describing the examples, we omit all mathematical details, these fall beyond the scope of this report. Several projects have motivated more theoretical research activities; these are carried out at the Eindhoven University of Technology, under the supervision of the author. A brief description of some of these research interests will be given in section 15.4, "Research activities in production management."

15.3 Projects in production management

In this section, we briefly describe a number of studies carried out in the field of production management in several Philips organizations. Mathematical details are omitted.

15.3.1 Production and inventory management in a Telecommunication industry

Figure 66 shows the logistics diagram of the production process of our first example, i.e. the assembly of large office exchanges for voice- and data-transmission. The production process of these exchanges (starting with the supply of components and ending with the installation of the exchange) can be divided in three important phases, separated by physical stock points. In the first phase, components and raw materials (electrical components, integrated circuits, cables, wood, etc.) are delivered by external suppliers. In the second phase the production of subassemblies takes place (cables prepared for connection, shelves and, in particular, printed circuit boards). In the third phase we find the final assembly, the functional tests, the packing and the expedition. Often, also transport and installation are included in this phase. Two warehouses exist: the component store (components and raw materials) and the commercial store (for all subassemblies). Finally, we note that some subassemblies are also delivered directly from outside suppliers.

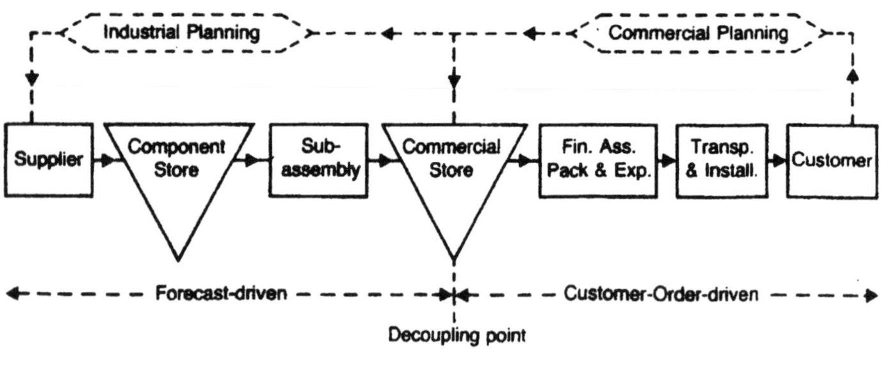

Figure 66. Logistic diagram of the exchange production process

The production planning department of the company works according to the following rule. Upon arival of an order, a **commercial delivery time** is agreed upon with the customer which is sufficient to incorporate the time needed to complete the third phase of the production process (i.e. final assembly, tests, packing, expedition and sometimes transport and installation). In our example, the total leadtime needed to complete this third phase varied from 2 to 4 months, depending on the size and the complexity of the exchange. Hence, upon arrival of a completely specified order, the necessary amount of all types of subassemblies (specified by the Equipment Survey) must be available, to make sure that the final assembly of the exchange can be started almost immediately.

It follows that purchase orders for components and production orders for subassemblies have to be released before customer orders that need the subassemblies have arrived (or before all details of a customer order, such as size, special features, etc. are completely specified). Therefore, production of subassemblies will be driven by **inventory control rules**, based on **forecasts** of the demand. The logistic performance of the production process depends heavily on the choice of these rules and in particular on the value of the parameters which determine the size of production orders, safety stock levels, etc. The third phase (final assembly, etc) is then completely **order-driven**. We also say that the **decoupling point** in the logistics chain is located between the subassembly and the final assembly phase, at the commercial store.

We have developed a so-called **Master Production Scheduling** (MPS) rule for the production of subassemblies, which served as the basis for a computer-based **Materials Requirements Planning** (MRP) program (for concepts like MPS and MRP, see Orlicky (1975)). This MPS was based on the concept of the **Economic Inventory Position** (see e.g. Silver and Peterson (1985)) of all subassemblies. This Inventory Position is formally defined as

```
Economic Inventory Position (subassemblies) = Inventory in commercial store
                                            + released production orders
                                              for subassemblies
                                            - subassemblies, committed to
                                              customer orders
```

Note that the work in process inventory in the subassembly phase is included in the economic inventory position, since also released production orders are counted.

It is important to realize that, in our example, the inventory positions of hundreds of different types of subassemblies have to be recorded. Since this requires a lot of data processing capacity and time, the inventory positions are updated only once in a fixed review period (instead of continuously). As a result, production orders for subassemblies are also released only once per review period.

The Master Production Scheduling rule for period t is based on the forecasted demand in the period (t,t + R + L), where R denotes the length of the review period and L the subassembly leadtime. It takes into account a **safety stock** factor, based on forecast errors of the demand, as well as **lotsize** considerations. Also, commonality of subassemblies was considered. We speak of a high **degree of commonality** if the same subassembly can be applied in a large variety of final exchanges; such a high degree of commonality requires a highly **modular product structure**, based on standardized subassemblies, of these final exchanges. The final MPS rule orders subassemblies such that the Economic Inventory Position of these subassemblies is returned to an order-up-to level S_t (based on the above mentioned forecasts) at the beginning of every review period. This rule is known to perform well in the case of nonstationary demand (see e.g. Silver and Peterson(1985)). A detailed description of the MPS rule can be found in De Kok and Zijm(1988).

Since the production planning procedure for subassemblies highly influences average commercial stocks and work in process inventory levels, we next built a simulation model to investigate the effects on these inventories of various actions, including:

- reduction of the subassembly leadtime L,

- better pre-information from the Sales Department to Production Management in order to reduce forecast errors and therebye safety stocks,

- standardization of subassemblies, resulting in a higher commonality degree and hence again lower safety stocks in the commercial store,

- reduction of both subassembly and final assembly leadtime, without changing the commercially agreed delivery time of exchanges. This may ultimately enable Production Management to shift the decoupling point to the component store, thus making the production of subassemblies customer order-driven. The absence of uncertainty, together with the leadtime reductions, cause a dramatic reduction in inventory levels.

The model enables decision makers to evaluate the effects of these actions in quantitative terms of FAV (Factory Accounting Value) and Turnover (the rate of annual sales divided by the average inventory levels, both expressed in terms of money). Hence, the model serves as a **Decision Support System**, helping Production Management to choose the proper mix of actions to be taken to improve the competitive position of the Telecommunication company by a severe cost reduction.

15.3.2 Design of an automated manufacturing system for transformers

For the production of certain new types of transformers, which perform several important functions in a TV set, one of Philips' factories has installed a number of highly productive and flexible manufacturing lines. An automated conveyor system moves products (one by one) on coded product carriers between different locations, except for some oven processes where transport is in batches. All production steps, handling and tests are mechanized or automated. Also, the transport of the products is under rigid computer control.

Before building such a system of production lines, it seems desirable to gain insight into the performance of a proposed design, to predict the effects of a large number of variations, or even to evaluate completely alternative designs. Performance must be understood in terms of productivity, throughput and waiting times, sensitivity of a line with respect to machine breakdowns (reliability), flexibility with respect to product mix and variability in demand, etc.

Let us highlight one design issue in more detail. In order to cope with uncertainty in the production process caused by machine breakdowns, buffers have to be situated at crucial points in order to achieve a satisfactory production rate. Absence of buffers would cause the whole line to stop each time a machine breaks down. However, since all transport functions are completely automated and all products to the system are coded, any buffer system must be integrated with the automatic conveyors. Spacial and financial considerations place a severe constraint on the size of buffers. The question now is how the effective productivity of a line depends on the location and the size of these, necessarily limited, buffers.

A simple example may illustrate what happens. Figure 67 depicts a line, consisting of only two machines and one finite buffer (of size K) in between. Both machines are unreliable. Figure 68 shows a sample path of the stochastic process representing the behavior of the buffer contents as a function of the machine conditions. Note that machine 1 may continue production, even if machine 2 fails, as long as the buffer is not completely filled; in the other case we say machine 1 is **blocked**. Also, machine 2 may continue production, even if machine 1 has failed, as long as the buffer is not empty; in the other case we say machine 2 is **starved**. If machine 2 is the faster one, then it can only work at a rate equal to the production rate of machine 1 in the case of both machines working and an empty buffer in between; we say that machine 2 is **slowed down**. It will be clear that the **effective capacity** of the line is severely influenced by the size of the buffer.

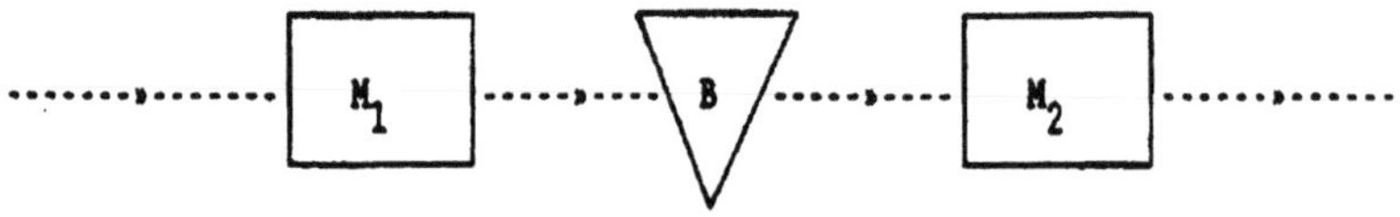

Figure 67. Two serial unreliable machines with a finite buffer

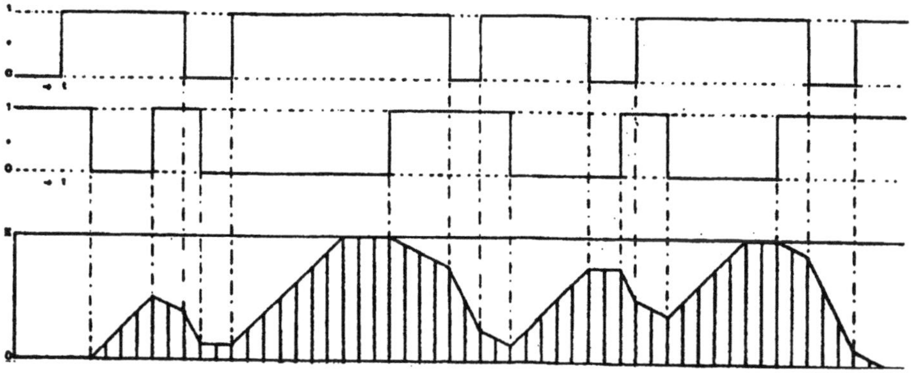

Figure 68. Buffer behavior corresponding to alterations in machine states. When both working, machine 2 is the faster one.

When lifetimes (times between breakdowns) and repair times are specified by probability functions of phase type (see e.g. Neuts(1981)), it is possible to calculate the effective capacity of such a line by solving a system of differential equations (using a fluid model approach) or by analyzing a specially structured Markov Chain (using a queueing-theoretical approach). See e.g. Wijngaard (1979) or Neuts (1981), ch. 5. Approximations for longer lines have been studied by several authors; a fluid model approach for this particular Philips case has been described by Wessels, Hontelez and Zijm(1986).

Returning to the transformer manufacturing lines, a natural decomposition appears to be possible. First, demand characteristics allows the allocation of each line to one particular family of products, within these families products can be manufactured alternately, without changeover times of machines. Hence, flexibility with respect to product mix is assured.

Furthermore, from an aggregate point of view, each line can be divided into three parts. In parts 1 and 3, products are moved one by one on unit product carriers, running through a set of small cycle time operations and tests; in the middle part where several oven processes take place, transport is in large batches (on multiple product carriers). Product handling between the unit product carriers and the multiple product carriers is performed by specially designed robots. Machines in part 1 and 3 are not completely reliable, the middle part is almost perfect. Between machines in part 1 only very small buffers (of two or three products at most) are allowed. The same holds for part 3. Between the three parts, hence in front of and immediately behind the oven processes, larger buffers are allowed, because of storage possibilities on multiple product carriers.

Using a fluid model approach, in combination with a simulation study, we evaluated a number of alternative designs and we developed a control rule for dispatching products to the line. In particular, the following results were obtained.

- The optimal sizes and locations of the buffers were determined. On the one hand these buffers had to be as small as possible, on the other hand they had to be large enough to guarantee a desired minimum throughput.

- The best performance of the line was achieved by giving it a "push-pull" character. Recall that, from an aggregate point of view, the line could be divided in three main parts. The first part had to **push** products into a buffer in front of the middle part, thereby guaranteeing input for the (expensive) ovens, even when, due to breakdowns, that first part is blocked temporarily. The latter part **pulls** products from the middle part - even when taking into account its disturbance rates - therebye preventing overflow or an excessive amount of work in process.

- For capacity balancing reasons, some operations should be located at a point parallel to the line.

- At the beginning of the line, a number of parallel machine units were situated along the conveyor system. Each machine unit has its own cycle time. Products visit only one of the parallel machine units. In order to avoid blocking on the conveyor system, caused by products operated at one machine (in line), a control rule has been developed for dispatching products to these machines.

Valuable insight with respect to the performance of the different alternatives has thus been obtained. The proposed model and the subsequent analysis appeared to be powerful instruments to support the design process of these type of production lines. These types of models are now frequently used, also after the realization of production lines, to assist management in making the right decisions when product or process changes need to be introduced. So it may properly be thought of as a management tool.

15.3.3 Shopfloor scheduling in a cable factory

Our third example concerns the production of mains leads, with attached plugs, in a cable factory. These leads are then distributed to other product divisions such as the Consumer Electronics Division (audio and video equipment), Domestic Appliances, etc. The leads are supplied on reels by another department (one reel may contain several kilometers of lead). The processing of leads with plugs takes place in three phases. In the first phase, the leads are cut to the right length (between 1.5 and 3 meters), stripped at the end, the end points are soldered, etc. In the second phase, the plastic plugs are attached by an injection moulding machine. In the third phase finally, certain specified leads are tested (only those leads for which a test certificate is required by the customer) and all leads are packed in boxes.

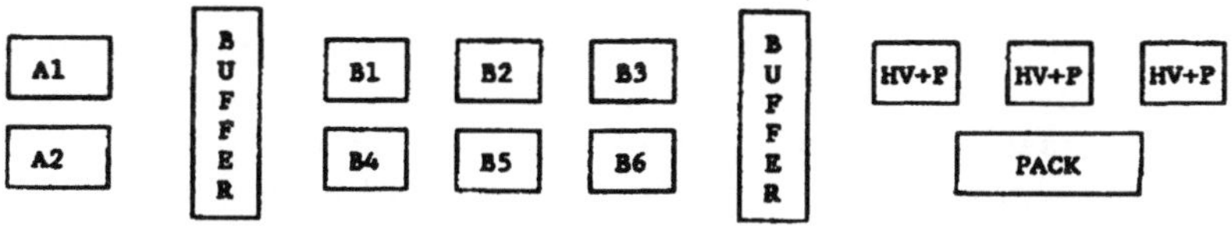

Figure 69. Production layout for mains leads manufacturing

For the first phase, two identical machines are available, for the second phase six identical machines. In the third phase, three high-voltage test machines are available, leads that pass through one of these three machines are packed immediately after testing. Another part of the daily production is packed directly.

Between each two phases buffers are available to store products temporarily, on the floor (in case no machine in the next phase is directly available for processing that particular kind of product). Buffers however have only limited capacity.

Weekly production plans are made for this mains leads department, taking into account due-dates (delivery dates which may be even within that week), the availability of material and in particular changeover times. Both the machines in the first and in the second phase are characterized by **sequence-dependent changeover times**, i.e. the changeover time depends both on the type of product just produced and the type of product to be produced. Since there exist about 200 different type numbers, **a matrix of changeover times** would require 40,000 entries (200*200). It appeared to be possible however to considerably reduce the number of space required by taking a closer look at the changeover times.

For the two machines in the first phase, the changeover times are built up as the sum of the times needed to perform changes in a number of tool settings. For each tool setting, there are only a few choices possible (at most four in our case). Each type number is characterized by a specified tool setting for each tool needed. Hence, a database which specifies the tool settings for all 200 product types, together with a small database which contains the times needed to perform changes in the tool settings, is sufficient to **calculate** for each group of part types, to be produced in the next week, say, the changeover time matrix when needed. The total changeover times are relatively large when compared with the time needed to produce 1000 leads (which is about half an hour, whereas the changeover times may vary from 8 minutes to at most one and a half hour). In the second phase, something similar happens. The only difference is in fact that, instead of the sum, the maximum of a number of times has to be taken. In the third phase, changeover times are negligable and can be ignored.

A Master Production Plan specifies the group of orders to be produced in the next week. Next, the shopfloor scheduler has to determine a schedule which specifies for each machine when precisely to produce a particular order. Orders vary in size from 2000 to 25000 leads. In one week, 40 to 50 orders have to be produced. Orders may be split among several machines in each phase. We were asked to develop a scheduling methodology that could be implemented on a small personal computer, to assist the shopfloor scheduler in evaluating different alternatives (under slightly different constraints) quickly and to enable him to quickly **reschedule** part of the set of orders in case of serious interruptions (e.g. rush orders or machine breakdowns). Before our involvement, the scheduling was done manually on a large planning board (actually a Gantt chart was constructed). A brief outline of the way we approached the problem is given below.

First, we determined in which phase the capacity limitations were most severe. In our case, the two machines in phase one constituted the bottleneck, partly caused by the fact that these machines suffered from more or less serious breakdowns. The next important observation was that the planning problem for one group of machines (in one phase) closely resembles a so-called **multiple travelling salesmen problem** with **time constraints** (cf. Lawler et. al.(1985)). In this problem, a group of salesmen (machines in our case) have to visit a set of cities (to process a set of orders), where the distance between city i and city j is given by c_{ij} (the changeover time from order i to order j is given by c_{ij}). For each city, earliest entry times s_i, visit durations p_i and latest departure times t_i are specified. In our scheduling problem, where cities are replaced by orders, these variables denote material avaibility dates (from a preceding phase or a preceding department), processing times and due-dates (or dates at which material must be available for a subsequent phase), respectively. The problem is then to minimize the total time needed to visit all cities exactly once (to process all orders), taking into account the time windows (material availability and due-dates), by specifying for each salesman (each machine) a particular **sequence** of cities (orders), to be visited (processed) in that order.

Although the original problem setting was slightly different, the main question was to produce as efficiently as possible hence to minimize the time lost to changeovers, which in a natural way constitutes the multiple traveling salesmen problem with time constraints for each phase. The only, not unimportant, difference with the usual traveling salesmen problem is that our scheduling problem is essentially **asymmetric** (i.e. a changeover time c_{ij} from type i to type j is in general not equal to c_{ji}). Since all known (heuristic) algorithms for traveling salesmen problems with time constraints are developed for the symmetric case, we had to develop new heuristics.

We developed a heuristic (based on local search techniques) for scheduling the two machines in the first phase (since this phase appeared to be the bottleneck). We omit mathematical details again. If not absolutely necessary (because of due-dates) orders are not split and processed on two machines in parallel, in order to avoid additional changeovers. In planning the first two machines, we loosely take into account the main changeover times in the second phase, in order to prevent arriving at sequences which would cause a very unsatisfying changeover pattern in the second phase.

The resulting schedules appeared to be a very serious improvement over the manually prepared schedules (an improvement of 20% was the rule rather than the exception). Even more important however was the fact that the runtime needed to arrive at a good schedule on a micro-computer was only a matter of seconds, thus enabling the shopfloor scheduler to use the method as an instrument in what-if simulations. All kinds of unforeseen events could be handled easily now by proposing new schedules, starting from the situation at which the event occurred (this constituted another constraint on our traveling salesmen problem formulation which however could be easily handled). In this way, a practical and easy to handle instrument was made available to the shopfloor controller, to enable him to develop production schedules within a few minutes, to evaluate whether a proposed Master Production Plan is feasible indeed and, if not, to evaluate alternatives, and to reschedule quickly in case of breakdowns or rush orders, a situation which could not be handled in a satisfactory way when planning manually.

15.4 Research activities in production management

The first responsibility when carrying out projects in factories, is to respond **properly** to the problems posed by management, within a reasonable time. Problems in factories or logistics organizations have to be solved adequately, but a reasonable trade-off has to be made between the efforts needed to replace a good solution by an optimal one (if possible) and the benefits that can be expected from a very minor improvement of an already good solution to for instance a production planning problem. Besides that, optimality is not always clearly defined in the often turbulent environments we are working in.

On the other hand, one often feels the need for a more basic theoretical understanding of certain problems, for instance because their appearance in many different places in many different forms justifies such a serious research investment, or simply because the importance of the area is recognized by top management. Another (very good) reason may be the personal interest of a researcher in the field. In our case, many projects led to research activities which are carried out at both Philips and the University of Technology in Eindhoven. In this section, we briefly indicate some of these activities, carried out at the Mathematical Department of the EUT, under the supervision of the author.

15.4.1 Global performance analysis of automated transport systems in factories

This problem area was motivated by several projects, among which the case described in section 15.3.2, "Design of an automated manufacturing system for transformers" and a study concerning the redesign of the transport system in a vacuum cleaner factory in the Netherlands. One may think of automated conveyor belts but also on so-called AGVS's (Automatic Guided Vehicle Systems) or railcart systems which both are often applied in Flexible Manufacturing Systems (see e.g. Ranky(1983) or Zijm(1987)). Queueing network models have provided valuable insights into the behavior of these systems (cf. Stecke and Suri(1986)). In our research, we concentrate on approximative network models, and in particular on issues such as traffic priority rules, integration with local or centralized buffer systems, etc.. A comprehensive description of the queueing analysis of the vacuum cleaner case can be found in Repkes and Zijm(1988).

15.4.2 Machine scheduling problems

Many machine scheduling problems can be classified as so-called **generalized flowshop scheduling problems**. All products have to pass through a sequence of machine banks, where each bank consists of one or a number of parallel machines. Products visit at most one machine in each bank (products may skip a bank). Machines in each bank may suffer from changeover times or breakdowns. Orders may be subject to release and due-dates, they may or may not be split in smaller lots, etc. Certain orders may have a higher priority than others. With respect to the product structure a certain family structure may be apparent.

Combinatorial optimization methods can be exploited to solve only rather small problems in manufacturing environments which are a very special case of the above sketched general situation (e.g a single parallel machine system or a simple flowshop with only one machine in each phase and no changeover times at all). In our research, we concentrate on approximation methods for more complex environments such as the generalized flowshop described above. The approach is based on decomposition methods, using combinatorial procedures for the smaller problems as our building blocks, and exploiting a rather sophisticated iterative aggregation procedure recently proposed by Adams et. al.(1988). A first attempt to solve a generalized flowshop scheduling problem without changeover times is described in Zijm and Nelissen (1988).

Another important research topic is the analysis of these scheduling procedures in a rolling planning environment (where the set of orders may change frequently) and the development of order acceptance procedures based on feasibility of the eventually resulting schedules.

15.4.3 Multi-echelon production/inventory control systems

Consider a logistic chain, consisting of a number of suppliers, a components warehouse, a factory (possibly to be split into a subassembly and a final assembly department), a central warehouse for final products, several local warehouses and finally retailers and market. To develop models which adequately describe the many complex interactions in such a chain still appears to be extremely difficult, despite the many attempts that have been made in the past. We first devoted our attention to a description of these interactions, both in terms of the Physical materials flow and in terms of the flow of **information** in such a system (including the role of a central planning department). Next, we concentrate on a comparison between a Base Stock Control System and a Manufacturing Resources Planning System (compare e.g. Vollmann et. al.(1984)). In the near future, we wish to include concepts such as Hierarchical Production Planning, commonality of components, size and location of safety stocks, flexibility in capacity and the like.

A central role in our models and analysis methods is played by the concept of **echelon stock**, developed by Clark and Scarf (1960) and, in our view, highly underestimated by both researchers and practitioners. The decomposion approach, proposed by Clark and Scarf, has been generalized to more complex environments, compared with the simple line system studied by these authors. A first review article will appear in the beginning of the next year (Langenhoff and Zijm (1989)).

15.4.4 Design and control of flexible assembly systems

About 90 % of the literature in the field of Flexible Manufacturing Systems concerns Machining Centres, more in particular in a Metal Cutting Environment. We have planned to study **assembly systems** and, as an example, we have taken the mounting of printed circuit boards, being one of the most widespread processes in electronics manufacturing. Problems we study include:

- Machine control problems. How to load components on a particular insertion machine? How to control so-called **pick-and-place** devices?
- How to spread a total amount of work over a number of parallel machines?
- What are preferrable production configurations? Line structures or assembly cells?
- What MHS (Material Handling System) should be chosen?
- How to estimate overall performance of different assembly configurations?

Right now, we concentrate on the last issue. We try to develop approximation methods, based on queueing theory, for difficult assembly structures. In particular, we study job-dependent parallel structures, i.e. a system of basically parallel machines, visited by jobs belonging to several different classes (types of printed circuit boards), where each job type can only visit a subset of the group of parallel machines. The subsets however may overlap which causes essential difficulties. Moreover, jobs may choose a queue according to a shortest queue principle, a very common control rule in practice which unfortunately leads to known hard problems in a queueing-theoretical sense. First results have been obtained, see Adan et. al.(1988).

15.5 References

[1] Adams, J., E. Balas and D. Zawack(1986), The shifting bottleneck procedure for job shop scheduling, Management Science Research Report No. MSRR-525, Carnegie Mellon University, Pittsburgh,

[2] Adan, I., J. Wessels and W.H.M. Zijm(1988), Queueing analysis in a flexible assembly system with a job-dependent parallel structure, to appear in: H. Schellhaas et. al. (eds.), Operations Research Proceedings 1988, Springer-Verlag, Heidelberg,

[3] Clark, A.J. and H. Scarf(1960), Optimal policies for a multiechelon inventory problem, Management Science 6, pp. 475-490,

[4] David, E.E.,Jr.(1984) , Renewing U.S. Mathematics: Critical Resource for the Future, Report of the National Research Council's Ad Hoc Committee on Resources for the Mathematical Sciences, Notices of the AMS 31, pp. 435-466,

[5] De Kok, A.G. and W.H.M. Zijm(1988), Production Planning and Inventory Management in a Telecommunication Industry, in: A. Chikan (ed.), Proceedings of the Fourth International Symposium on Inventories, Elsevier, New York,

[6] Forrester, J.W.(1961), Industrial Dynamics, The M.I.T. Press, Cambridge, Massachusetts,

[7] Holt, C.C., F. Modigliani, J.F. Muth and H.A. Simon(1960), Planning Production, Inventories, and Work Force, Prentice-Hall, Inc., Englewood Cliffs, N.J.,

[8] Jackson, J.R.(1963), Jobshop-like Queueing systems, Management Science 10, pp. 131-142,

[9] Langenhoff, L. and W.H.M. Zijm(1989), Multi-echelon production/inventory systems with assembly and distribution structures, to appear,

[10] Lawler, E.L., J.K. Lenstra, A.H.G. Rinnooy Kan and D.B. Shmoys (eds.)(1985), The Traveling Salesman Problem: a guided tour of Combinatorial Optimization, Wiley, New York,

[11] Neuts, M.F.(1981), Matrix-geometric solutions in stochastic models, an algorithmic approach, John Hopkins University Press, Baltimore,

[12] Orlicky, J.A.(1975), Materials Requirements Planning, McGraw-Hill, New York,

[13] Ranky, P.G.(1983), The Design and Operation of Flexible Manufacturing Systems, IFS Publications, Kempston, Bedford, U.K.,

[14] Repkes, J.P. and W.H.M. Zijm(1988), Analysis of automatic transport systems in an assembly factory, to appear in: H. Schellhaas et. al. (eds.), Operations Research Proceedings 1988, Springer-Verlag, Heidelberg,

[15] Silver, E.A. and R. Peterson(1985), Decision Systems for Inventory Management and Production Planning, 2nd. ed., Wiley, New York,

[16] Stecke, K. and R. Suri (eds.) (1986), Flexible Manufacturing Systems, Operations Research models and applications, Elsevier, Amsterdam,

[17] Vollman, Th. E., W.L. Berry and D.C. Whybark (1984), Manufacturing Planning and Control Systems, Dow Jones-Irwin, Homewood, Illinois,

[18] Wessels, J., J. Hontelez and W.H.M. Zijm (1986), Eine Produktionslinie mit unzuverlassigen Komponenten (in German), presented at the 15th DGOR - meeting, Ulm, FRG,

[19] Wijngaard, J.(1979), The effect of interstage buffer storage on the output of two unreliable production units in series with different production rates, AIIE Transactions 11, pp. 42-47,

[20] Zijm, W.H.M. (1987) , Flexible Manufacturing Systems: background, examples and models, in: H. Schellhaas et. al. (eds.), Operations Research Proceedings 1987, Springer-Verlag, Heidelberg, pp. 142-161,

[21] Zijm, W.H.M. and E.H.B.L. Nelissen (1988), Scheduling a Flexible Machining Centre, to appear in A. Chikan (ed.), Proceedings of

the Fifth International Symposium on Inventories, Elsevier, New York.

Part Two: Contributions from Italian Industries

16. Mathematics as a Fundamental Tool in the Analysis, Design and Simulation of the Behaviour of Space Systems

F. Bevilacqua - C. Maccone

AERITALIA S.A.I.p.A. - Space Systems Group

Corso Marche 41 - 10146 Torino - ITALY

16.1 Abstract

The Space Systems Group was created at the Aeritalia facilities in Turin, Italy, about 25 years ago. Within the Space Systems Group, the Special Studies Unit has ever since been engaged in the most important space programs fostered by the European Space Agency (ESA) as well as by the Italian National Space Plan (PSN). This activity centers on the analysis, design and simulation of the behaviour of space systems, requiring a large amount of sophisticated calculations. Several software packages are currently used by us for mathematical modelling: DCAP, SINDA, ESATAN, PATRAN, in addition to other minor programs to be run on personal computers such as MATHCAD, MATLAB and MUMATH. A review of these programs, with a few critical remarks, is presented in this paper. An opportunity is thusly provided for discussing the interplay between the mathematical community, responsible for breaking up new pathways in the solution of problems, and the industrial world, that these solutions must put to work in real time.

16.2 Introduction

The present paper points out the rôle of the mathematical software currently available at the Space Systems Group of Aeritalia for the analysis, design and simulation of space systems. Part of this software (i.e. DCAP) is directly produced by Aeritalia itself, whereas other programs have been purchased by the company for the internal use.

16.3 DCAP (Dynamics and Control Analysis Package)

The increasing complexity of the spacecraft configurations requires extensive analytical and experimental studies in order to ensure the overall confidence in the performance of the systems. This applies in particular to the payload deployment and/or retrieval mechanisms, to the antennas pointing and shape-control, and to the large flexible structures. Most important, and potentially most difficult, among these studies are the analysis of the dynamic response and prediction of stability. In particular, these studies are required during the design of the configuration and control, where the most complex characteristic of the system plays a central rôle.

In the above context, the European Space Agency (ESA) is pursuing the development of a multipurpose Dynamic and Control Analysis software Package (DCAP), for which Aeritalia Space Systems Group bears the responsibility of actual software producer. In practice, DCAP consists of some thirty thousand Fortran statements, providing a general-purpose modelling capability for the dynamic simulation and control analysis of actively controlled advanced spacecraft. The user is thus assisted in performing the analysis of complex, multibody, flexible structures and their associated control systems. The tasks related to this analysis may be classified as follows:

1. **Time history simulation.**
   DCAP offers the capability of two types of time history simulation: linear and non-linear. The linear simulation is based on a linear model of the structure obtained via the linearization of the equations of motion in the neighbourhood of a certain instant. This linearization aims at reducing the cost associated with the time history simulation of complex systems. On the other hand, the non-linear time simulation, is the real heart of the DCAP system, in that the equations of motion are essentially non-linear. This is because DCAP is oriented towards the simulation of multi-body structures in an arbitrary topology. That is, the individual bodies can be connected together in an arbitrary fashion by means of 'hinges'. Each of the individual bodies can be either rigid or flexible. In addition, the flexible bodies can be described in a number of ways: modal models, finite elements models, simple beams, etc.. The hinges are arbitrary point connections between bodies, with six degrees of

freedom per hinge. Each of these degrees of freedom can be constrained to zero motion, or constrained to be a specified function of time, or be unconstrained. Also, the structure has attached sensors and actuators for interfacing with a control system, as now explained.

2. **Control systems synthesis.**
   The control system synthesis utilizes a variety of programs to help the linear system analysis and synthesis. The capability starts with a linearization program that linearizes the equations of the multi-body structure about some nominal operation point. The control system can then be linear, continuous or discrete, and multi-rate or user-defined. The entire structure may be specified to operate in an earth environment complete with the effects of gravity and of gravity gradient. The user can also obtain the frequency responses of the system in open or closed loop configurations. These frequency responses can be plotted in anyone of the Nyquist, Nichols, Bode and root-locus format typical of the design of the classical control systems. As a further aid to the control synthesis, DCAP offers LQG (Linear Quadratic Gaussian) programs for synthesizing a full state feedback controller.

When to use DCAP.

Since DCAP offers a full range of simulation and design capabilities, it can be used throughout the full life of a space program, from early feasibility studies and preliminary design to detailed performance assessment studies. In other words, DCAP provides the user with the capability of simulating less sophisticated models during the initial design phases; whereas, in the subsequent design phases, it allowes the user to increase progress. The benefit to the user is that DCAP does not become obsolete, and the user doesn't have to learn a new simulation system while the program progresses.

The DCAP multibody formalism.

An accurate dynamic modelling of the spacecraft is the basic prerequisite to any successful structure and control design, as well as to the achievement of high performances. In general, a spacecraft is today a complex system of dynamically constrained bodies, some of which are rigid, some of which are flexible or very flexible. As for the constraints, these may include different types of bounds that suppress certain degrees of freedom, whereas other constraints may be left free, or may be variously controlled. In general, large variations in the geometry of the space system are to be expected during its full operational time. The nature of the constraints too can be variable in time: for instance a very special case is that of a space system during the rendez-vous and docking maneuvre: the two parts are free up to the contact, and then get constrained (in a rather special way) while contacting. Also, large antennas in space, their deployment, pointing and control are further important cases requiring an accurate design of the structure, dynamics and control.

In view of the complexity of the system to be modelled, it is quite useful to have a clearly defined method of reducing the complex problem to an ordered set of smaller, formally repeatitive, standard coupled problems. This is just what the 'multibody formulation' does. The multibody formulation consists in dividing the system into a set of distinct 'bodies', and in formulating the equations of motion for each body independently of the other ones. In these equations, next to all other known inputs, a set of unknown 'constraint loads' is included to take into account the effect of the adjacent interacting bodies. By adding the 'constraint equations' to these equations, it is possible to solve the system for both the status variables of the bodies and the unknown constraint loads. The multibody approach is apparently nothing special, but actually represents a convenient bridge between two rather conflicting conceptual methodologies: the 'energy (or Lagrangian) methods', that eliminate all the constraints by the choice of suitable 'generalized coordinates', and the 'force (or Newtonian) methods' which, on the contrary, leave the constraint loads as unknown. In the DCAP multibody approach, the Lagrangian method is followed within each single body, while the 'force' concept is used in the practice for coupling different bodies. The model of the single body can be formulated in a standard form, and also the constraints, for a very wide class of problems, can be put in a standard form. The consequence is that the equations of motion of this system are standard, and that the user does not have to worry about writing them. The program cares about this; at the same time, the Lagranging formulation, kept within each body, guarantees computationally economic handling of the single body problem.

Among the state variables of the system, the constraint loads are often a required output of the simulation. When this is not the case, however, it is also possible to think of a reduction of the computational load, by eliminating these variables form the equation system before, or periodically along, its integration. This technique saves extra computational time.

The DCAP multibody formulation consists in modelling each body of the system as an 'elastic body'. This can actually be a continuous elastic member as well as an assembly of rigid and elastic parts. It rests in the hands of the analyst whether to split the overall system in larger or smaller subsystems; with many small bodies the user will probably input less elaborated data, while for larger and more complex bodies the information to be provided needs a more sophisticated elaboration. The single body information to be provided is in fact not only about the geometry and the mass, but about the elastic behaviour too. For the latter, DCAP uses the rather general approach of expressing the elastic displacement as

$$\eta(P) = \phi^T(P)\,\xi(t)$$

where $\phi^T(P)$ is a matrix of 'modal functions' and ξ a vector of 'modal amplitudes'.

The 'modal functions' are in general 'admissible' functions and are defined by the user; in particular, they can be actual vibration modes of the body. The latter (as well as their frequencies and other useful data such as the mass matrix etc.) can be provided by specialized structural programs, (ASKA, NASTRAN etc.) when the modelling is of a sophisticated nature. One of the most appealing features of DCAP is just the interfacing of structural and dynamic software, that is the capability of an integrated structure-dynamics-design. Better, this capability is seen to extend to the thermal aspects too if only one remembers that a structural program is a nice way to introduce in the system thermal deflection equivalent loads, in the form of static thermal deflection/shapes, and even thermal interactions so important in the large antenna design. DCAP is also well equipped for simulating other environmental loads, as for instance those arising from gravity gradient, or the apparent forces in non-inertial reference frames such as an orbiting frame.

Control Analysis and Synthesis.

The 'spacecraft' that has been modelled according to the preceding paragraph is now the 'plant' to be controlled by an appropriate control system. In general, it is a complicated, large, non-linear plant whose inputs are the forces (both the disturbances and the control actions at given stations) and whose outputs are the status variables (positions and velocities) at some selected stations. DCAP provides a suitable way for prescribing the input/output stations of the system by means of the so-called 'sensor points'. At such points a frame rigidly connected to the structure is defined, whose position and velocity is given in output and/or where the input forces are referred to for a rather general set of commonly used sensor and actuators.

A very important preliminary step in designing the control system is the assessment of the 'open loop transfer function'. This is an essentially linear concept, and can be applied to our non-linear plant only if a linearization process (for instance about an 'equilibrium' state) is preliminarly performed. This is accomplished in DCAP by means of a numerical linearization procedure. A plant matrix A is thus generated, which represents the linearized plant; this matrix can be subject to eigenanalysis and frequency response. The results can be presented in Bode, Nyquist or Nichols form. The A plant, defined above, does not actually include any part of the sensor and the actuator behaviour. In the hypothesis, however, that the input to the sensors was related to the state variables through an algebraic relationship (and the force input are so related to the actuator signals) these relationship are linearized together with the overall system; moreover the matrix A is transformed, by a similarity transformation, in such a way to replace some of the original state variables with the actual input-output signals variables. In this way meaningful transfer functions are generated. Besides the linearization capability, DCAP provides also means for the direct input of already linear systems, for both analysis and/or simulation purposes. On any linear system, the program provides means to perform:

i) system augmentation by pre-and-post signal filtering. By this operation, the system, inclusive of the actual sensor and actuator response, is modelled in 'modal' uncoupled form;

ii) modal again evaluation, where the weight of each mode on the various elements of the relevant transfer function matrices are evaluated, so that a 'model' reduction can possibly by performed on this rationale;

iii) root locus analysis.

The described analysis provide suitable means for designing the control system on a 'classical theory' basis. Modern control can however be implemented through the LQG design facility incorporated in DCAP Release 6. Both continuous and sampled control systems can be designed.

16.4 SINDA (Systems Improved Numerical Differencing Analyzer)

The SINDA program was created in the early seventies from the CINDA-3G program, or Chrysler Improved Numerical Differencing Analyzer for Third Generation computers. Essentially, both CINDA-3G and SINDA are suited for solving lumped parameter representations of physical problems governed by diffusion-type equations. In particular, SINDA was originally designed as a general thermal analyzer accepting resistor-capacitor (R-C) network representations; however, with due attention to units and thermally oriented peculiarities, SINDA will also accept R-C networks representing other types of systems.

To let the reader get a clearer picture of how SINDA works, we shall consider the analogy between the discharge of an electrical capacitor and the equivalent thermal system. The voltage across the capacitor's plates corresponds to the difference in temperature between any two assigned reference points belonging to our thermal system. And, finally, just as an easy differential equation governing the transient phase of the capacitor's discharge can be written and solved, so may be written and numerically solved the not-so-easy differential equations governing our thermal system.

SINDA consists of two main pieces: the processor and the library. The processor is a program accepting problems written in the SINDA language and converting them to the FORTRAN language. The SINDA library consists of many pre-written FORTRAN subroutines performing a variety of commonly needed actions and thusly reducing the programming effort to solve a given problem. These routines are fully compatible with the FORTRAN routines produced by the processor from the user's input.

16.5 ESATAN (European Space Agency Thermal Analyzer)

ESATAN is today the standard computer program of the European Space Agency for the analysis of any spacecraft thermal behaviour. It was developed at the Mechanical Engineering Laboratory of the General Electric Corporation under a contract with ESTEC (the European Space Technology Centre) as a replacement of the SINDA program for ESTEC and its subcontractors. ESATAN is written in ANSI Standard Fortran 77 for machine portability, and has been available since 1985.

It is enlightening to compare the similarities and differences between SINDA and ESATAN by reviewing the story of how ESATAN was originated from SINDA. The need for the development of ESATAN was identified by ESA during a review of their analysis methods for spacecraft thermal control purposes. This had concluded that, while the network modelling technique employed by SINDA whose satisfactory, SINDA itself had significant shortcomings in several areas. In particular, these included the checking of models for the correctness of purpose, the validation of modifications to existing models, and the merging of several independently developed models of system components to represent the complete system. In addition, there was a need to improve upon the numerical solution algorithms contained in SINDA. Finally, it was concluded that the incorporation of these extra feature into SINDA was not feasible, because the existing software structure did not lend itself to such extensions. The decision was thus taken to develop a new program to replace SINDA which would include all of these additional features. However, the fundamental approach of SINDA, consisting of separating the topology of the thermal model from the solutions performed on it, was still retained for ESATAN.

Thus, the basic aim of ESATAN is the prediction of temperature distribution in engineering components and systems using the thermal network analysis components. It enables the user to specify his problem by means of thermal network quantities such as nodes, conductances and material properties, together with the sequences of solutions required to obtain the required steady-state or transient temperature distributions. ESATAN also has an interactive editor capability allowing changes to be made to previously defined thermal network models with automatic checking of their and consistency. Finally, ESATAN uses a preprocessor to read interpret the user data; then the Fortran program is compiled and linked with a library of functional modules performing various tasks such as network solution and interpolation.

Althoug ESATAN is much more flexible than SINDA, it must be realized that it is, in general, three to five times slower too. This is basically due to the actions of the preprocessor, performing long files conversions. Therefore, the use of SINDA in the practice, has only in part be subseded by that of ESATAN. In fact it all depends on how involved the thermal model to analyze is. In case of simple models, the use of SINDA may still be preferable to save time and money, whereas, on the contrary, ESATAN reveals its full power in case of large and complicated structures.

16.6 PATRAN

PATRAN has now become the basic tool for inputting and visualizing the geometry of space structures.

PATRAN is a product of PDA, a software and engineering company focussing on composite materials and finite element analysis.

PATRAN Plus consists of five basic modules, plus gateway utilities. Four modules are contained as standard feature in the basic PATRAN executable program and one module is independent. These modules fulfill the following tasks: P/SOLID deals with the advanced geometry modelling, P/FEM with the finite element modelling, P/POST with the results evaluation, P/IMAGE with the imaging, and P/PLOT with x-y engineering plotting. The following application modules may also be added to expand the system capabilities: P/COMPOSITE for composite design and analysis, P/MECHANISM for mechanism design, P/STRESS for linear statics and dynamics analysis, and P/THERMAL for thermal analysis. Finally, the following interfaces with standard applications programs are available also: ABAQUS, ADINA, ANSYS, COSMIC-NASTRAN, IGES, MARC, MOLDFLOW, MSC-NASTRAN, SINDA, and VDA.

The gateway utilities of PATRAN Plus give users the freedom to define their own environment, while application interfaces bring the 'world' of mechanical computer-aided engineering into PATRAN. Virtually every major finite element code, as well as many important computer aided drafting and manufacturing software packages, can be tightly linked into the PATRAN interface system.

Finally, all data can be converted into a form compatible to SINDA for the solution of the structural problems, and the PATRAN postprocessor completes the procedure by creating the pictures on the screen.

16.7 Some Useful Mathematical Programs for Personal Computers

MATHCAD was produced in the mid-eighties by the American, Mathsoft Inc., and is a very useful numerical solver. It can quickly deal with such issues as resolution of linear and non-linear simultaneous equations, computation of definite integrals, simulations involving random numbers, and immediate display of graphs of any kind.

MATLAB is somehow similar to MATHCAD, but is more powerful in dealing with matrix operations. In particular, it can find the eigenvalues and eigenvectors of a given matrix, and show the various steps leading to the final result. The graphic capabilities of MATLAB are good too, and they prove especially useful in signal processing, inasmuch as MATLAB can compute and display Fast Fourier Transforms, Convolutions, etc.. Created by MathWorks Inc.

in the mid eighties, MATLAB proves to be a useful aid to both scientists and engineers.

Finally, MUMATH is a typical example of a symbolic mathematics program. It formally performs derivatives, integrals (whenever that is possible in a closed form), Fourier series, finite sums and products, trigonometric expressions, and is able to solve both algebraic equations (up to the fourth degree), and differential equations. Written in a language (called MUSIMP) derived from LISP, MUMATH includes a useful set of lectures providing a thorough introduction to the subject of artificial intelligence.

17. The Numerical Methods In Mechanical Components Peak Stress Analysis: Some Industrial Applications.

P.P. Strona

Centro Ricerche Fiat

17.1 Abstract

The peak stress analysis in mechanical components is very important in the Automotive Industry, both from the correct design and the fatigue life prediction point of view.

The approximation of the peak stresses by numerical methods is a quite difficult task.
Because of the shortcomings of classical Finite Element Method based on the "h" approach, particularly when solid elements must be utilized in the models, a strong research effort in the world has been done to improve the related analysis capabilities: Boundary Element Method and Finite Element Method based on the "p" approach have been developed and are applied now in the Automotive Industry.

Besides it is important to develop not only the codes which implement the best algorithms, but also to build the complete systems able to solve the whole problem of the analysis, including the automatized generation of the models and the synthesis of the most significant results, so that the designers could utilize good numerical methods in a very efficient and user-friendly way.

In the paper, after some general consideration and a synthesis of the most important advantages and shortcomings of the above methods, some examples of significant Automotive applications concerning spur and helical gears and truck frames with holes are shown.

17.2 General Considerations

In the Automotive Industry, in the last years, the evolution of the way of designing new components and structures makes more and more important the numerical methods.
The trends can be synthetized essentially in two main aspects:

> The designers try to define the best solutions of their projects by defining and modifying iteratively more and more realistic mathematical models.
>
> The experimental testing has some new different goals than in the past, particularly the validation of the numerical models and the check of the final design configurations.

The integrated development of numerical methods, of corresponding software, of hardware tools, particularly vector and parallel computers, make possible this evolution whose more significant general goals consist in a greater reliability and in a time design reduction for new vehicles.

Generally two kinds of applications are developed in the Automotive Industry:

> Higly sophisticated problems, for example crash analysis, technological processes simulation like metalforming and forging, fluidodynamics and combustion problems.
>
> Daily Technical offices calculations, particularly components and structures stress analysis, essential step in the verification of the different successive design configurations.

In the paper two applications where the peak stress analysis is a fundamental step for a good and reliable component design are shown:

The first [1-2-3-4-5-6-7] concerns the Gear Teeth Stress Analysis and the second [8] concerns the Truck Frame Stress Analysis near Holes.
The related activities developed in Fiat Research Center are summarized focusing particularly the improvements obtained in the utilization of the numerical methods.

17.3 Gear Teeth Stress Analysis

The determination of the peak stresses in the fillet of spur and helical gears is one of the most important steps in the gear teeth design procedures both from the correct design and fatigue life prediction point of view.

Classical methods are based on some semi-empirical formulas: the problem is simplified by some drastical hypotheses and the load sharing among the engaged teeth can be taken into account with several noticeable difficulties.

Today the evolution of the numerical methods allows the designers to deal with this problem by a more realistic, reliable and powerfull sophisticated models.
The Finite Element Method (FEM) in both its "h" and "p" versions and the Boundary Element Method (BEM) are available for the analysis.

According to the studies performed in Fiat Research Center during the last years on this topic, it is possible now to summarize the most important advantages and shortcomings of the above methods when they are applied to the gear teeth peak stress analysis as follows:

BEM is very suitable and efficient in this kind of problems because it requires the mesh definition only in the body surface, a line in 2D models and a surface in 3D solid models; the transition from coarse to refined mesh regions can be easy to do and the accuracy in the peak stress maximum values is good because the mesh is defined just on the body surface itself.

The more critical shortcoming is the structure of the matrix generated according to the formulation which is fully-populated and unsymmetric: a substructures approach should be perhaps adopted when the analysis of large 3D solid models was required so that a banded matrix could be obtained.

Tha BEM advantages and shortcomings are practically the corresponding shortcomings and advantages of the "h" version of FEM.

The other possible choice is the "p" version of FEM.

While with h-FEM the results accuracy depends on the number of elements,with p-FEM it depends on the polinomial level of the element shape functions which are hierarchical.
Consequently when a p-FEM model is generated, the mesh can be defined once from the geometrical point of view and the same geometrical mesh can be analyzed with several different degrees of freedom distributions related to the polinomial degree which has been selected in each element.

Besides the mesh can be generally coarse in these components modeling because it depends mostly on the compliance with the actual geometry of the teeth and on the allowable distorsion degree.
At last The p-FEM formulation is suitable to an "Adaptive Procedure" to check the results accuracy.

According to the above considerations BEM has been selected for the analysis of spur gears by 2D models.
Figure 70 on page 235 and Figure 71 on page 236 show respectively the h-FEM and BEM models of the same pinion tooth and Figure 72 on page 236 shows a comparison between the corresponding obtained peak stress results.

Instead p-FEM has been selected for the analysis of helical and spur gears by 3D solid models.
Figure 73 on page 237 and Figure 74 on page 238 show the models generated respectively for spur and helical gears analysis.
All the meshing teeth are included in the models.
The torque is applied as an equivalent uniformely distributed load at the bottom surface of the pinion model.
The contact between pinion and gear teeth is simulated by a multipoint constraint condition: the non-linear effects are not considered because only the accuracy in the load sharing among teeth is requested for a good valuation of the fillet stress values.

Figure 75 on page 239 and Figure 76 on page 240 show an example of output of the analysis.
It is very important to focuse that two special purpose software systems have been developed respectively for BEM and p-FEM analysis.

BEM code is an in-house developed code while p-FEM code is a commercial one developed by ISMES, whose name is "FIESTA".

Pre-processors programs allow to generate the complete respective models automatically at all starting from very few data necessary to define the problem.
Post-processors programs allow to collect the most significant results useful for the designing needs.

17.4 Truck Frames Holes Stress Analysis

In truck frames, which are generally subjected to high level loads, it is necessary to make holes of different size to realize the cables lay-out or to join several components to the tie-down points.
Therefore the life prediction of the frames requires an accurate valuation of the peak stresse around the holes.
BEM has been selected as a basis of an analysis system developed to investigate this problem.

The numerical procedure has been tested from the experimental point of view, then it has been applied for a parametric investigation.

The tested truck frame is shown in Figure 77 on page 241.
Figure 78 on page 242 and Figure 79 on page 242 show respectively the geometrical configuration and the corresponding BEM model: the problem consists in studying a plane stress condition of a plate with bending load in its plane and with two eccentric holes.

The convergence results have been obtained analyzing meshes with different number of elements along the holes boundary.
In Figure 80 on page 243 the numerical and experimental results comparison is displayed.

17.5 Conclusions

Some general considerations about the design procedures evolution in the Automotive Industry have been explained.
Two industrial applications concerning the peak stress values analysis have been shown.

The related developed activities and the obtained results allow us to observe that the numerical methods and their improvement are more and more important in the design procedures to build reliable and realistic models and to get a better accuracy in the results. Besides it is also important to allow the designers to utilize the above highly sophisticated numerical techniques in a very efficient and user-friendly way by implementing several "ad hoc" software systems with pre and post-processing programs suitable to solve as completely as possible the corresponding problems.

17.6 References

[1] M.Alemanni,S.Bertoglio,P.P.Strona BEM in gear teeth stress analysis: Comparison with classical methods Int.Symposium on Gearing and Power Transmissions Tokyo,1981.

[2] M.Alemanni,S.Bertoglio,P.P.Strona Applicazione del Metodo degli Elementi al Contorno all'analisi di denti di ingranaggio diritti lungo l'arco di ingranamento XI Convegno Nazionale AIAS. Torino 1983.

[3] M.Alemanni,P.P.Strona Il Metodo degli Elementi al Contorno nel calcolo dei denti di ingranaggio: criteri di modellazione ATA, Ingegneria Automotoristica, Novembre 1985

[4] P.P.Strona Il Metodo degli Elementi al Contorno nelle procedure di Ottimizzazione di Trasmissioni ATA, Ingegneria Automotoristica, Novembre 1985

[5] A.Pidello,F.A.Raffa,P.P.Strona The p-version of the finite element method in gear teeth stress analysis T.K.S. Murthy, C.A. Brebbia Eds. Structural Design and Crashworthiness of Automobiles Springer Verlag, 1987

[6] C.Baret,A.Pidello,P.P.Strona An Integrated system for the stress analysis of spur and helical gears Int.Symposium on "Use of Supercomputers in the European Automotive Industry, Torino 1988

[7] C.Baret,A.Pidello,P.P.Strona The 3D stress analysis of spur and helical gears: an integrated system based on the p-FEM approach AGMA Fall Technical Meeting,New Orleans,1988

[8] F.M.Cura',F.A.Raffa,O.Rehmann,G.Ronchiato,P.P.Strona The Boundary Element Approach in the evaluation of Stress Intensity Factors of Plane Structures with Holes Int.Conference BEM87, Stuttgart, September 1987.

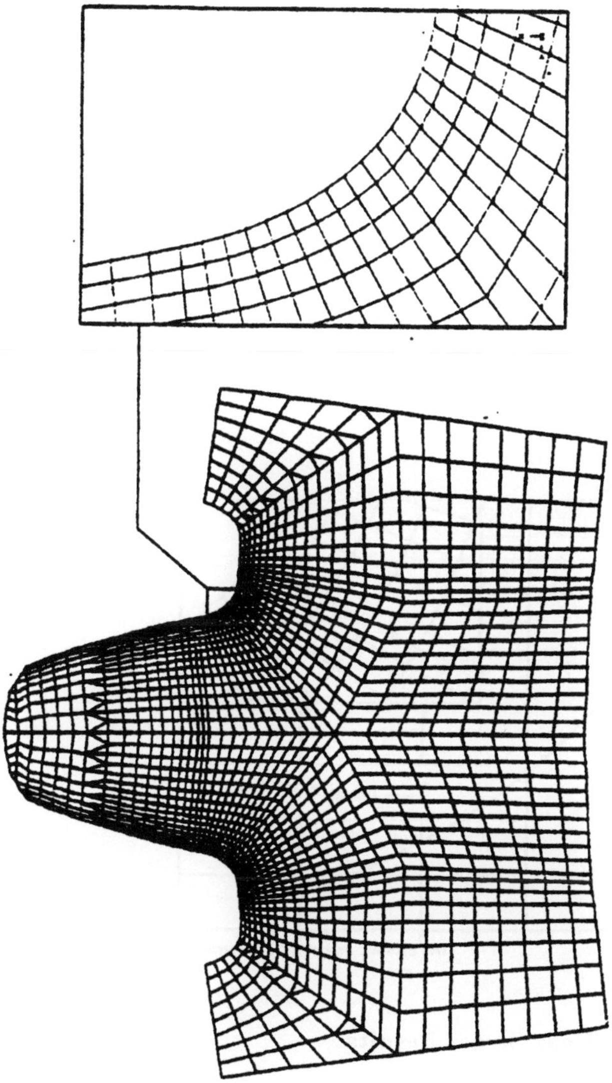

Figure 70. Spur Gear NASTRAN Model (FEM)

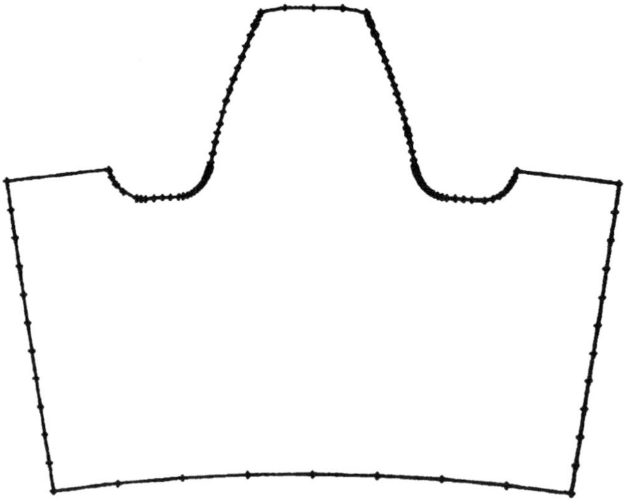

Figure 71. Spur Gear BEM Model

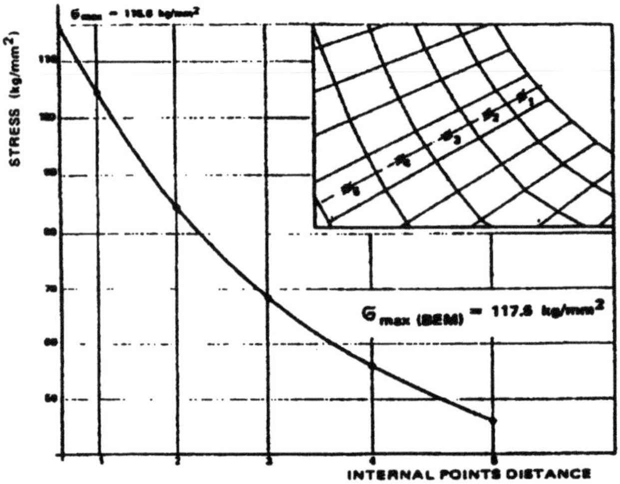

Figure 72. Spur Gear Stress Results

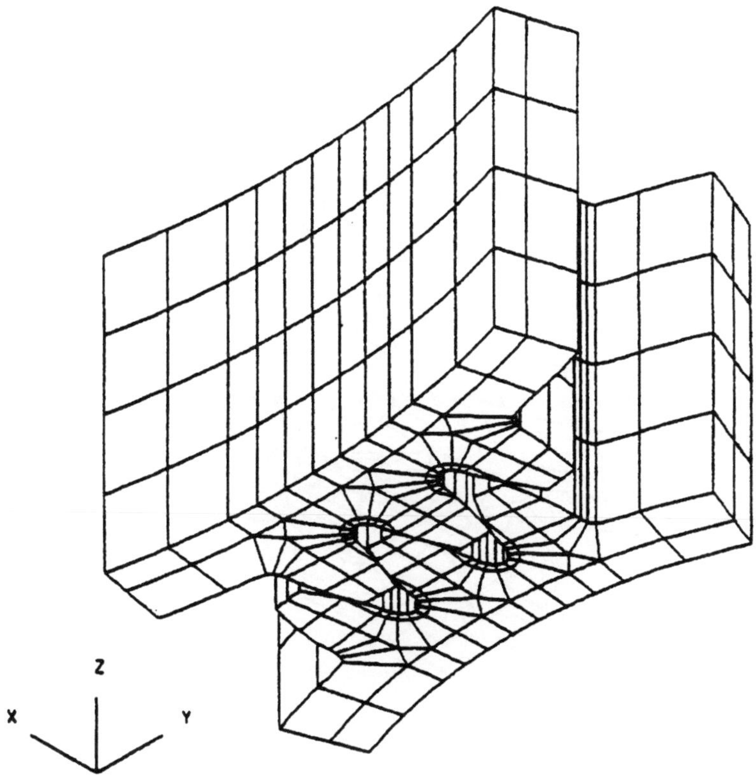

Figure 73. Spur Gear Meshing Teeth Model

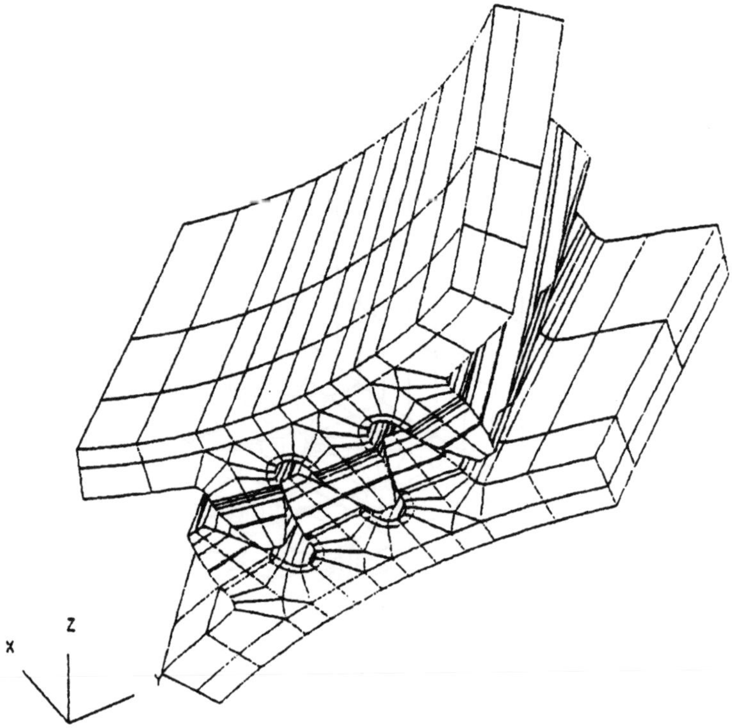

Figure 74. Helical Gear Meshing Teeth Model

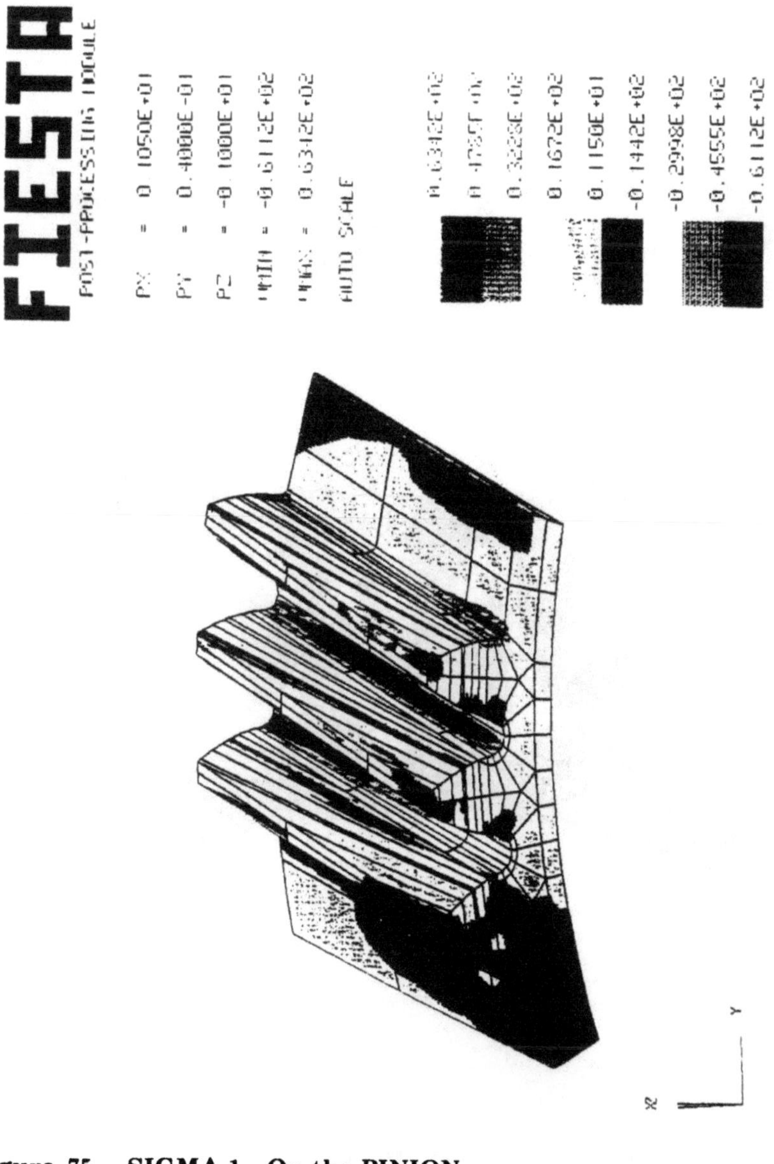

Figure 75. SIGMA 1 - On the PINION

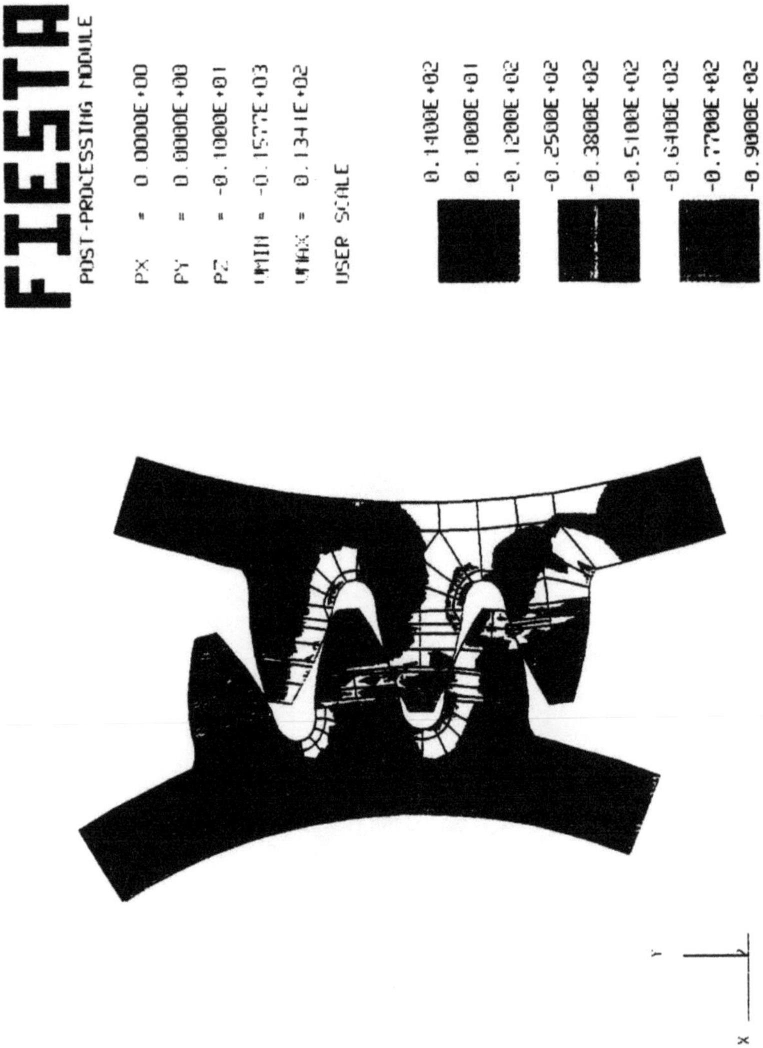

Figure 76. SIGMA 3 - 3D Model - Helical Teeth

Figure 77. General View of Truck Frame and Test Equipment

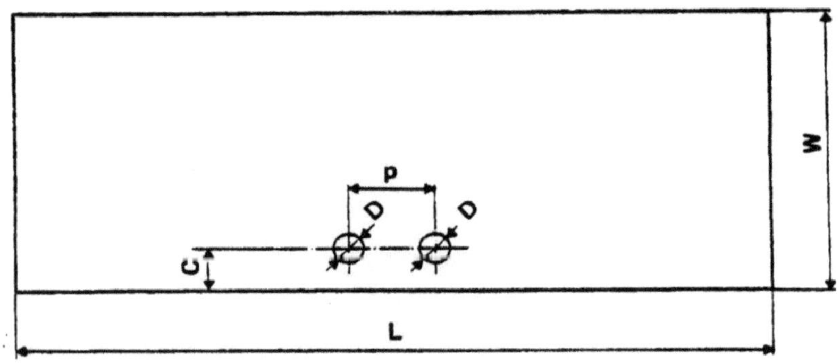

Figure 78. Geometrical Configuration

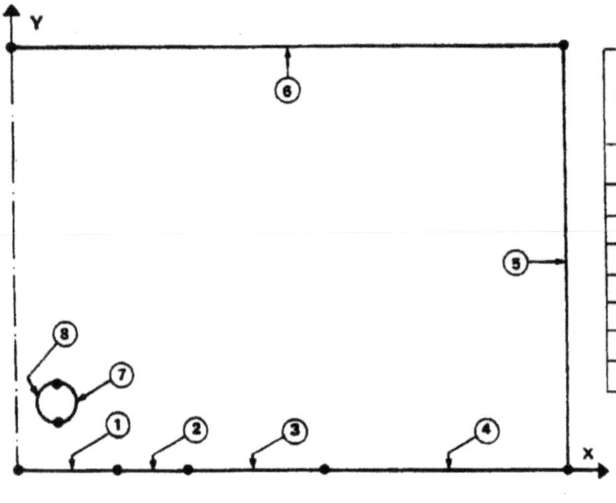

Region	N of Boundary Elements
1	12
2	4
3	4
4	7
5	10
6	15
7	7 ÷ 50
8	7 ÷ 50

Figure 79. BEM Model

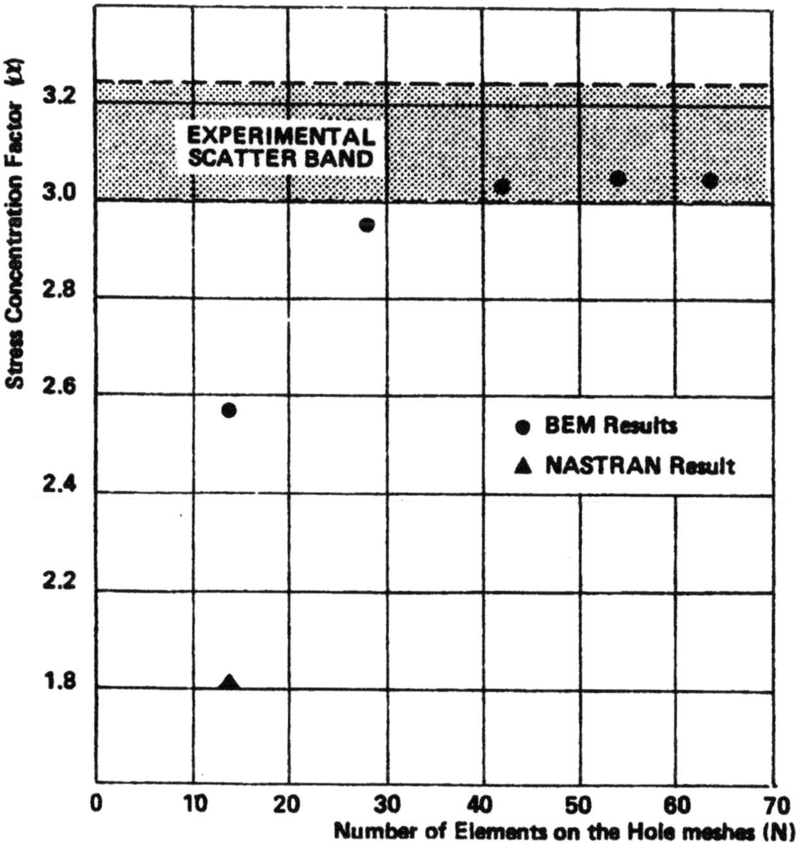

Figure 80. Comparision of Experimental and Numerical Results (p/D = 3.67)

18. Simulation Programs in Chemical Engineering. The Choice of Eniricerche

M. Bistolfi, G. Soave

ENIRICERCHE

18.1 Introduction

Simulation programs are nowadays indispensable tools for both engineering and research companies in the chemical field.

Such programs allow users to reproduce the behavior of a single piece of equipment (pump, heat exchanger, distillation column, ...), several units or whole chemical plants.

These programs can perform physical and thermodynamic properties calculations (density, viscosity, specific heat, ...) and mass and energy balances; they account for phase changes (evaporation, condensation, phase splitting) and chemical reactions; they reproduce unit operation equipment.

In order to obtain such information, these programs include in their structure a data bank about the most common chemical compounds and their basic properties; a number of routines that calculate the properties and the conditions of the streams considered; and modules to simulate equipment that use the calculation routines to obtain the desired results.

In addition to these calculation sections, you need input and output sections to establish the conditions at which the simulation has to be performed (temperatures, pressions, compositions of the streams), and to produce the output of the results.

To simulate complex networks, including many recycle streams thus requiring solution by iteration, you must also have a section assuring the convergence of the algorithm for the whole plant; in addition you could optimize the output of the plant from the energy saving point of view, for example; in these case you have to cope with "flowsheeting" rather than "simulation" problems.

All these considerations regard steady-state simulation or flowsheeting programs, but very recently also "dynamic" simulators have been developed: they can manage start-up and shut-down operations, introducing in their logic the time variable; their use is still unusual.

Many steady-state simulators have been developed and commercialized during recent years and they are commonly used for design purposes; but the more complete these programs are, the more complex in general is the approach for the unexperienced user.

For a researcher or for a research & development engineer, who needs not use it as frequently as a process engineer, this complexity should represent a great disadvantage.

Thus the choice of our company, ENIRICERCHE, was to develop an interactive simulation program (PERSIA) that uses a series of panels and menus, to retrieve information from the data bank and to assign the conditions for the calculations.

This interactive feature is also useful to display the convergence of a single calculation (i.e. distillation column) or of a network.

Other important characteristics of PERSIA are its modularity, which permits easy implementation of new routines, and vapor-liquid equilibrium calculation methods; some of them (Soave - Redlich - Kwong, UNIWAALS) developed by G. Soave of ENIRICERCHE, are among the most recent and accurate for polar and non polar mixtures.

18.2 Program Structure

Simulation programs like PERSIA contain a number of different modules, either to store the data or to perform calculations.

In our program the most important modules are:

- A compound properties data bank
- A calculation methods data bank
- A memory input file
- A working streams file

While the first two file simply store useful data to retrieve, the others are created and modified by the program: the input file contains the description of the equipment to simulate, while the streams file contains the characteristics of all the streams used and produced by the program; the latter file is also checked to assure the convergence in case of iterations.

As program blocks we have:

- An input block
- A calculation block
- An output block

The calculation block is composed of a number of methods to calculate thermodynamic and physical properties of the streams (equations of state, activity coefficient relationships, empirical correlations, ...), and by the unit operation modules which use those methods.

18.3 Interactive Facilities

PERSIA is an interactive program which has a series of self-explaining panels and menus to facilitate entering data to the unexperienced user; each of these facilities are rapidly managed using predefined function keys of the terminal.

With the first panel of the program you can choose to retrieve or modify information about the properties of the compounds; otherwise it will appear the panel with the list of the compounds and the units to use during the simulation.

18.4 Type of Calculations

With the unit operations menu one can perform with PERSIA simple vapor-liquid equilibrium calculations, create new streams, mix or split them, simulate the behavior of equipment or of network.

If you choose equilibrium calculations, the program will ask you for the composition of the input streams then for the type of calculation you want.

If you choose any unit operation, it will ask you for the operating conditions (temperatures, pressures, type of equipment, number of stages, ...).

18.5 Calculation Methods

Another important choice you have is about the proper thermodynamic method to select for the simulation, for the equilibrium constants and for the enthalpy. Some of these methods have their own constants to use, others ask the user for specific parameters.

Among these methods, the equations of state called RKS and UNIWAALS habe been developed by G. Soave of ENIRICERCHE and are among the most reliable predictive methods for polar and non polar substances.

18.6 Conclusions

Simulation programs are common and useful tools in chemical engineering.

PERSIA is a user-friendly simulation program developed for research purposes.

PERSIA modular structure facilities future implementations.

PERSIA predictive methods allow reliable calculations in process design.

19. FEM Modelling with Marc. Application Studies

A. Cella, G. Noberasco

ESPRI - MARC Srl - GENOVA

19.1 Abstract

The mathematical modelling with the Finite Element Method (FEM) has raised within the last decade to become an industrial standard. That is due to the large variety of applications [1], and to the robustness of the computational algorithms.

FEM analysis to-day covers the following major application areas :

- structural and stress analysis;
- thermal and thermo-mechanical analysis;
- electric and magnetic fields;
- fluid dynamics;
- metalforming.

A general purpose FEM program, MARC, covers all of them [2].

Of particular interest are the high-tech applications associated with the behaviour of the new materials (composites, elastomers, polymers), or with the industrial forming of metal components (extrusion, forging, deep stretching).

Also the analysis of the ultimate behaviour of structural components is in growing demand: instability, collapse, fracture mechanics. Similarly for the analysis of high stress concentrations: shocks, impact, contact.

The program MARC is distributed worldwide by MARC Analysis, since 1972. It is strongly present in high-tech areas.

In Italy MARC has been in use for over a decade. Since 1981 ESPRI-MARC distributes, installs, supports and provides training for the MARC Software.

The Engineering Division of ESPRI-MARC has worked for over 30.000 man ours of consulting on special projects, in 8 different industrial sectors.

19.2 Finite Element Modelling

The Finite Element Method has been raised, by extensive research over the last two decades, into an industrial standard for the numerical simulation of many physical problems :

- stress analysis of mechanical components;
- thermal and thermomechanical analysis;
- distribution of electric and of magnetic fields;
- fluid flows;
- mechanical behaviour of materials;
- metalforming.

Amongst the many contributors, Prof. O.C. Zienkiewicz [1] had a prime role in exploring these new application areas.

Numerical simulation, due to the high efficiency compared with lab experiments, brings to an industrial product the value of extensive quality assurance: improved design, controlled performance, better manufacturing.

To-day most of the high cost components are analyzed by FEM codes: spacecraft, aircraft, ships, cars, nuclear and thermal power plant components. But the trend to achieve improved design is spreading to less expensive components: industrial vehicles, home appliances, industrial plants.

The trend is due both to the rapid decrease in the costs of computation and to the growing competition for higher quality and safety.

The new materials: composites, polymers, elastomers require advanced FEM analysis, because the material behaviour is no longer isotropic and no longer linearly elastic. Of particular interest is the assessment of the ultimate resistance of composite materials.

Finally, there is the emerging sector of manufacturing technology. FEM can now simulate and monitor numerically the forming processes of metals : deep sheet drawing, extrusion, casting, forming, rolling, coining, etc. These kinds of FEM analysis are highly demanding in computer power; vectorial machines are needed.

19.3 The MARC program

MARC is a general purpose FEM code [2] established in the industrial community since 1971. It is used worldwide.

MARC covers the whole spectrum of known FEM applications, including all the high-tech applications described here.

In the following we survey the major features of MARC.

MARC has four comprehensive libraries.

- Procedure Library
- Element Library
- Material Library
- Function Library

You may combine almost any number of options from each of the four libraries, and, consequently, you can solve virtually any structural mechanics or thermal problem.

- **Procedure Library**: This includes all the analysis types available in MARC:

Linear elastic	standard linear finite element analysis superposition of multiple load cases Fourier (non-axisymmetric) analysis of linear axisymmetric bodies.
Substructuring	multilevel, linear, quasistatic.

Nonlinear	automatic load incrementation elastoplastic scaling to first yield isotropic, kinematic and combined hardening large deformation/finite strain total and updated Lagrangian approaches buckling/collapse, linear/nonlinear creep buckling postbuckling, with adaptive load step rigid plastic flow, Eulerian, metalforming creep, with adaptive time step viscoelastic state equations (Kelvin model) hereditary integrals (generalized Maxwell or generalized Kelvin-Voigt model) thermorheologically simple behaviour viscoplastic, modified creep option to include plasticity effects contact/friction, automatic convergence.
Fracture mechanics	linear/nonlinear brittle/ductile J-integral evaluation cracking
Dynamics	modal analysis/eigenvalue extraction inverse power sweep method Lanczos method transient response modal superposition direct integration: Newmark-beta method Houbolt method central difference method harmonic response spectrum response time-stepping, linear/nonlinear adaptive time-stepping algorithm.
Heat transfer	steady-state and transient analyses, conduction, linear/nonlinear convection/radiation boundary conditions internal heat generation latent heat/phase changes adaptive time steps.

Hydrodynamic bearings	lubrification problems pressure distribution and mass flow.
Joule heating	coupled electric flow with heat transfer.
Fluid/structure interaction	incompressible and inviscid fluid.
Thermo-mechanical	quasi-coupled thermally driven stress analysis fully coupled thermo-mechanical analysis solved by a staggered scheme heat generated by plastic deformation and friction effects large displacement effects on thermal boundary conditions automated contact/friction capability.
Change of state	transient thermal analysis with change of phase and volume associated stress analysis with plasticity and residual stresses.

- **Element Library**

  MARC has a library of approximately 100 elements, covering the FEM geometries, the basic element shapes (triangles, quadrilaters), linear and quadratic isoparametric approximation, full and reduced integration.

- **Material Library**: This includes more than 40 different material models:

Linear elastic	isotropic, orthotropic and anisotropic (properties may be temperature dependent).
Composites	laminated plates and shells isotropic, orthotropic, or anisotropic layers elastic or elastic-plastic behaviour arbitrary material orientation definition with respect to any element edge with respect to global Cartesian axes with respect to a user-defined axis or through user subroutines relative ply angle for each layer

	multiple failure criteria maximum stress maximum strain Tsai-Wu Hill Hoffman, or user defined.
Hypoelastic	nonlinear elastic (reversible).
Elastomers	nonlinear elastic, incompressible Mooney-Rivlin model (allows large strains).
Elastic-plastic	Prandtl-Reuss flow rule user-defined non-associative flow law von Mises yield criterion isotropic, kinematic or combined hardening strain hardening (or softening) as a function of strain rate and temperature temperature dependence of yield stress and work hardening slopes isotropic, orthotropic and anisotropic Hill's anisotropic plasticity.
Cyclic plasticity	isotropic, kinematic, combined hardening.
Creep	deviatoric or volumetric (swelling) strains piecewise linear or exponential forms for rate of equivalent creep strain temperature dependence Oak Ridge National Lab. model, combines creep, plasticity and cyclic loadings.
Viscoelasticity	Maxwell and Kelvin models combined Kelvin-Voigt and Maxwell models hereditary integrals of strain histories with both small and large strain formulations thermorheologically simple behaviour isotropic or anisotropic material.
Polymers	thermorheologically simple behaviour.
Viscoplasticity	combining plasticity and the Maxwell model of plasticity.

Soils	yield surfaces as a function of hydrostatic stress linear or parabolic Mohr-Coulomb law.
Concrete	low-tension cracking crushing surfaces rebars.

- **Function Library** This includes a variety of MARC mesh generation and post- processing plotting options, kinematic constraints, loads, bandwidth optimization, rezoning, in-core and out-of-core solution, user subroutines, restart, output of postape, selective point, error analysis, etc. Only loads and constraints are summarized below.

Load and constraints	mechanical loads, concentrated, distributed, centrifugal, volumetric forces thermal loads, initial temperatures read from a postape produced from a thermal analysis, or from data cards initial stresses and initial plastic strains kinematic constraints transformation of degrees of freedom elastic foundation tying (multipoint constraints or MPC's) boundary conditions in user-defined axes springs and gaps, with and without friction.

19.4 References

[1] O.C. Zienkiewicz The Finite Element Method Mc Graw Hill (1978)

[2] MARC User Manuals, Vol. A,B,C,D,E MARC Analysis Research Corporation Palo Alto - California - USA

20. Dynamic Modelling of Industrial Manipulators

P. Bologna, G. Magnani

FIAR S.p.A.

Via I° Maggio 41, 20021 BOLLATE (MI)

20.1 Abstract

The object of this paper is to present the dynamic simulation of the FIAR industrial manipulator. The simulation is based on the continuous system modular simulation code LEGO.

LEGO permits to breakdown the system into subsystem (blocks) and to model each one independently. The system model is obtained assembling the blocks models. LEGO solves automatically differential and non-differential implicit equations (loops) arising from interconnecting subsystem variables. These features are of major importance in industrial manipulator simulation for they allow to bypass elegantly the difficulties inherent in mechanical chain modelling. The model coding activity is greatly reduced as well.

20.2 Introduction

Mathematical modelling of the manipulator dynamics is of major importance in the design and development of advanced manipulator controllers.
Dynamic models are used during the design phase to simulate the control system behavior and to evaluate its performances and stability. As manipulators are usually complex, multivariable, nonlinear dynamic systems, the more advanced control schemes implement a manipulator model to compensate for non linearities and interactions.

The simulation models determine the motion of the manipulator from a set of applied forces or torques (forward dynamics problem);

the control models, on the opposite, determine the forces required to produce a desired motion (inverse dynamics problem).

For the modelling purposes a manipulator can be viewed as made up of three main parts: the motors, the transmission gears and the mechanical chain which consists of bodies (links) connected by one degree-of-freedom (DOF)joints.

Closed form analytical models of common industrial manipulator mechanical chains (6 o 7 DOF) are hard to obtain for their equations are made up of a great numbers of terms. The simpler case to consider is when the chain links are arranged in sequence and do not form closed loops (open chains). In this case, numerical solutions have been attained, based on Newton-Euler equations [1], [2]. The inverse problem is the easier one to solve, for being known the motion forces and torques can be recursively computed from the last link (the gripper) to the first one (the base) [1]. The forward problem is solved by [2] exploiting the inverse solution. Analytical models have been attained using symbolic computation [3], [4].

With regard to closed mechanical chains neither analytical nor numerical general for formulations have been established, to our knowledge.

We present in this paper a dynamic model of the FIAR redundant manipulator. We have modelled the Dexter chain, which contains closed loops, by exploiting the features of a continuous process modular simulation code.
This code, named LEGO, appears to be highly suitable for robotics applications, and more specifically to model mechanical chains, in that:

- it allows a modular structure for the process model;
- it automatically solves algebraic implicit equations (algebraic loops).

Thus, even if acceleration and (generalized) forces acting on a link are instantaneously related to those acting on all the other links of a chain it is possible to obtain the model of the chain by modelling each link independent of the other and then assembling the models of the single links according to the LEGO rules.

This paper is organized as follows: section 20.3, "The LEGO package" contains a short outline of the LEGO code, section 20.4, "The Dexter manipulator structure" illustrates the Dexter manipulator structure and section 20.5, "Mathematical model of the manipulator" describes the model of the manipulator.

20.3 The LEGO package

LEGO is a continuous system modular simulation code which has been developed at the Automation and Computing Research Center of ENEL (the Italian Electric Power Board). Its more common applications are in the field of Electric power plant simulation and control design [5]. While a detailed description of the package can be found in [6], the main LEGO features can be outlined as follows.

LEGO consists of two main parts:

(i) The Modules Library (ML), made up of self consistent mathematical models (modules) each one representing a (fairly general) subsystem or component or element of a continuous dynamic process;

(ii) The Computation Master program (CMP), devoted to the handling and coordination of all the computation activities and, in particular, to the merging of the relevant elements of the ML and the solving of large-scale systems of non-linear algebraic and differential equations.

A module essentially consists of a system of non linear algebraic and differential equations of the following general form:

$$\dot{x} = f(x, y, u, q), \quad n \text{ equations}$$

$$O = g(x, y, u, q), \quad p \text{ equations}$$

where $u \in R^m$ is the vector of the *input variables*, $x \in R^n$ and $y \in R^p$ are the vectors of the differential, (state) and algebraic *output variables* respectively, and $q \in R^r$ is the vector of geometrical-physical parameters (design data).

Modules can be assembled, according to the LEGO rules, to build up the model of a complex process. The process is split into subsystems (blocks) each one modelled by a module in ML. A name (label) is given to each input and output variable of each block. Physical and mathematical interconnections between two variables of two blocks are established assigning the same label to them. Loops, also algebraic, obtained by interconnecting variables are supported by LEGO.

Two advantages related to modular simulation may be highlighted:

- the problem of modelling large scale systems can be solved by modelling systems of a smaller scale;
- large scale systems are frequently modular, in that they are composed of many elements of relatively few type, so that a limited number of module is needed. This is indeed the case of manipulator simulation.

As far as numerical aspects are concerned, LEGO's main features are:

- discretization of the differential equations according to the trapezoidal rule and simultaneous resolution of both the discretized and algebraic equations according to the Newton-Raphson iterative method;
- use of sparse matrix techniques;
- steady state computation which makes it possible to select the unknown variables among the input and output variables and among the uncertain constant parameters (obviously the number of the unknown variable must be equal to the number of equations);
- inherent resolution of implicit equations.

The code also includes user interface functions for the input of process data and the analysis of simulation results.

20.4 The Dexter manipulator structure

Dexter is a gantry, redundant (7 DOF) assembly manipulator (Figure 81).

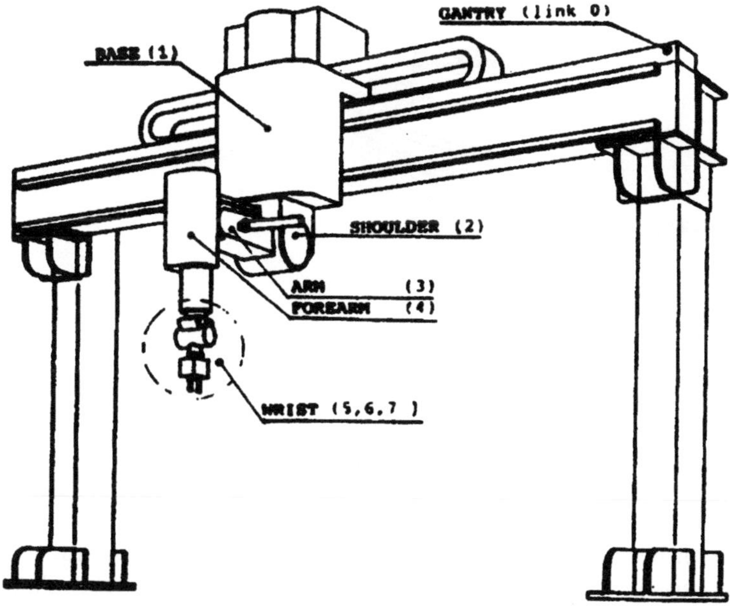

Figure 81. A Dexter sketch

A number of the Dexter mechanical chain features should be highlighted since they affect the dynamic modelling:

- the arm and forearm are connected as a parallelogram mechanism (Figure 82); this feature requires to model a closed chain.
- the arm is equipped with a spring-based mechanism for balancing the gravity forces;
- there are interactions among the movement of the wrist links: the three motors driving the wrist links are fixed to the forearm and their motion is transmitted to the links by means of coaxial shafts and gears enclosed in the forearm and in the wrist structure.

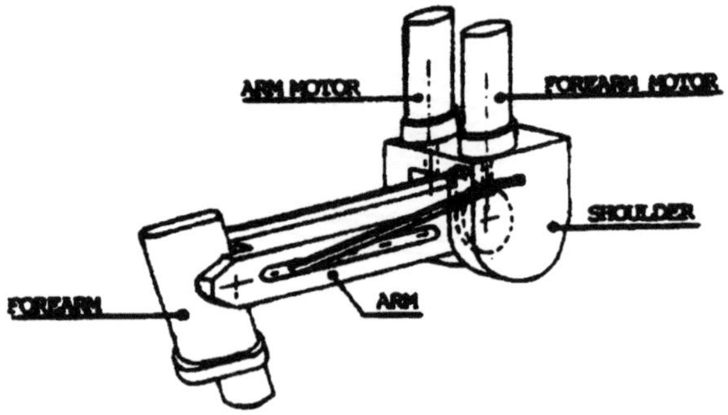

Figure 82. Arm and forearm parallelogram mechanism

20.5 Mathematical model of the manipulator

From the modular modelling point of view, a manipulator consists of three main components: the mechanical chain, the motors and the transmission gear trains. The mechanical chain, in its turn, basically consists of rigid bodies, the links, connected to each other by one DOF joint.

Thus three basic modules are required to model a general open chain manipulator: the LINK, the MOTOR and the GEAR-box modules.

20.5.1 The module LINK

This module represents a rigid body connected to two other bodies by rotational joints as the generic link of an open mechanical chain (Figure 83).

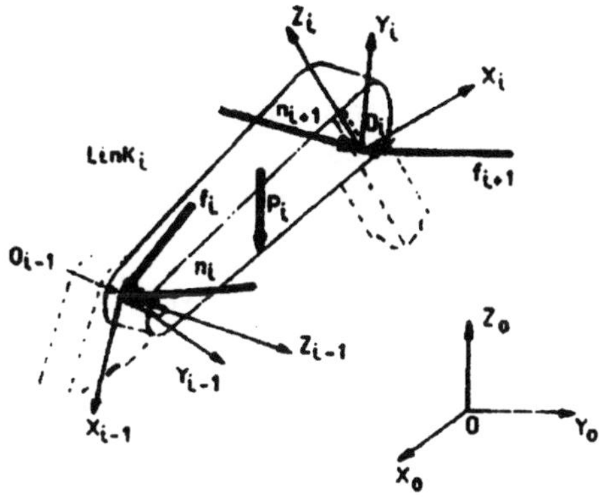

Figure 83. Forces and moments exerted on and from link i

According to the rigid body hypothesis, the link model can be based on the Newton-Euler equations as presented in [1], or, in a revised form in [7]:

$$w_i = w_{i-1} + z_{i-1}\dot{\theta}_i$$

$$\dot{w}_i = \dot{w}_{i-1} + z_{i-1}\ddot{\theta}_i + w_{i-1} \times z_{i-1}\dot{\theta}_i$$

$$\ddot{p}_i = \dot{w}_i \times p_i^* + w_i \times (w_i \times p_i^*) + \ddot{p}_{i-1}$$

$$\ddot{r}_i = w_i \times (w_i \times r_i^*) + \dot{w}_i \times r_i^* + \ddot{p}_i$$

$$F_i = m_i\ddot{r}_i + P_i$$

$$N_i = I_i\dot{w}_i + w_i \times (I_i w_i)$$

where

z_{i-1} axis of motion of joint i;
θ_i joint variable;
w_i , $\dot{w}_i$ absolute angular velocity and acceleration of link (and frame) i ;
p_i , p_i^* positions of the origin of coordinates (x_i ,y_i ,z_i) which is attached to link i with respect to the origin of link 0 and link $i-1$ coordinates, respectively;
r_i , r_i^* positions of the center of mass of link i with respect to the origin of link 0 and link i, respectively;

F_i , N_i total external force and torque on link i;
m_i mass of link i;
P_i gravity force: $P_i = m_i * [0,\ 0,\ 9.8\ m/s^2]^T$;
I_i inertia matrix of link i about its center of mass referred to the base coordinate axes. As I_i depends on the orientation of link i, we use the following expression for N_i (see [1]):

$$N_i = ({}^iA_0 * J_i * {}^0A_i) * \dot{w}_i + w_i \times [({}^iA_0 * J_i * {}^0A_i) * w_i]$$

where

J_i inertia matrix of link i about its center of mass referred to its own coordinates (x_i ,y_i ,z_i). We suppose J_i is a diagonal matrix, a common assumption (e.g. [8]);

${}^iA_0$ a 3x3 matrix of rotation which trasforms any vector with reference to frame i (x_i ,y_i ,z_i) to a coordinate system whose coordinate are parallel to (x_0 ,y_0 ,z_0) but its origin is the same as that of frame i;

${}^0A_i$ $= ({}^iA_0)^{-1}$.

The recursive equations for the forces and moments exerted on link i by link $i-1$ are:

$$f_i = F_i + f_{i+1}$$

$$n_i = n_{i+1} + N_i + (p_i^* + r_i^*) \times F_i + p_i^* \times f_{i+1}$$

$$T_i = z_{i-1}\, n_i + T_{fl}$$

where

f_i , n_i force and the moment exerted on link i by link $i-1$;

T_i input motor torque at joint i;

T_{fl} link friction coefficient.

To properly define the module, we also have to select the input and output variables. It is natural to assume as input (known) variables:

- the motion (angular velocity w_{i-1} and acceleration $\dot{w}_{i-1}$ and linear acceleration $\ddot{p}_{i-1}$) of frame $i-1$;
- the force f_{i+1} and the moment n_{i+1};
- the gravity force P_i;
- the motor torque T_i;

and as output (computed) variables:

- the motion (angular velocity w_i and acceleration $\dot{w}_i$ and linear acceleration $\ddot{p}_i$) of frame i;
- the force f_i and the moment n_i.

It is easy to see that that the model of a multilink open chain can be obtained as a cascade of blocks, each one modelled by the LINK module where the output variables of one block become the input variables of the previous one (f_i, n_i),and of the next one (the motion of frame i).

20.5.2 The module MOTOR

This module describes the electrical,mechanical and thermal behavior of a permanent magnet DC motor. According to [9] ,the equations describing the motor behavior are:

$$\begin{aligned}
L * dI/dt &= V - R * I - K_e * \Omega \\
J * d\Omega/dt &= K_t * I - T_{fm} - T_m \\
T_r &= K_t * I \\
d\theta_a/dt &= (R * I^2 + T_f \Omega - G_{ah} *(\theta_a - \theta_h) - G_{aa} * (\theta_a - \theta_e))/C_a \\
d\theta_h/dt &= (G_{ah} *(\theta_a - \theta_h) - G_{ha} * (\theta_h - \theta_e))/C_h
\end{aligned}$$

where

Ω	motor shaft velocity;
V, I	armature voltage, current;
J	moment of inertia of the motor shaft;
L, R	inductance, resistance of the armature winding;
K_e, K_t	motor voltage, torque constant;
T_{fm}, T_m	motor friction, motor output torque;
T_r	motor constraint reaction;
θ_e	average ambient temperature;
θ_h, θ_a	housing and armature mean temperature;
C_h, C_a	heat capacity of the housing and armature;

G_{ah} , G_{aa} , G_{ha} thermal conductance between armature and housing, armature and air, housing and air, respectively.

The armature resistance R and the motor constants K_e , K_t are described as linear functions of the armature temperature θ_a . The armature voltage and the angular velocity and acceleration are the module input variables, while the torques T_m and T_r, the current I and the temperatures θ_a and θ_h are the output variables.

20.5.3 The module GEAR

This module describes the dynamic behavior of the gear train transmitting the motion from motor shaft to joint.

The mathematical model of the gear train takes into account the inertia of the gears and the friction losses. As torsional flexibility is neglected, the motor shaft angular velocity Ω is related to the joint angular velocity $\dot{\theta}$ only through the gear ratio τ.

Under this assumption the equations of the module are:

$$\begin{aligned} J_t * d\Omega/dt &= T_m - \tau * T - T_{fg} \\ \tau * \Omega &= \dot{\theta} \\ \tau * d\Omega/dt &= \ddot{\theta} \end{aligned}$$

where J_t denotes the moment of inertia of the gear train, T_m and T are the motor and load torques respectively, T_{fg} is the gear friction torque. It is worth noticing that, when excited,torsional flexibility can give rise to torsional resonance. The servomotor bandwidth has to be limited to a frequency equal to a maximum of half the resonance frequency [10].

The input variables of the module are the joint angular velocity and acceleration and the motor torque T_m, while the output variables are the motor shaft angular velocity and acceleration and the load torque T (T being an input for the module LINK).

20.5.4 Friction

Friction has to be adequately modelled as it is of major importance in precision servo control [11], [12]. Different friction models proposed in the literature are discussed in [11].

The friction model we have chosen is shown in Figure 84.

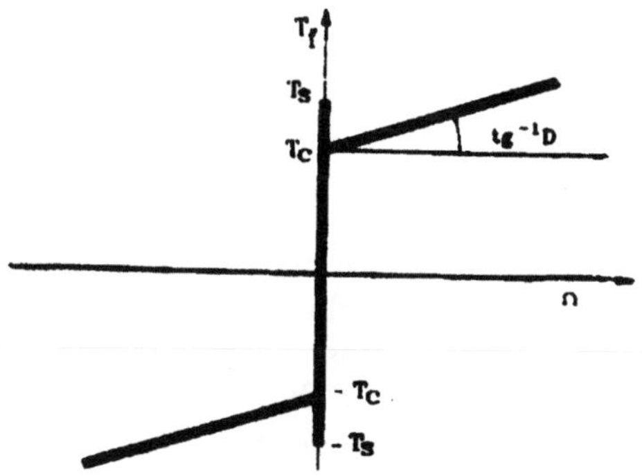

Figure 84. Friction model

The total friction torque (or force) T_f includes Coulomb friction T_c, viscous friction $D^*\Omega$ and stiction T_s.
Further details are in [13] where there are also the identification of uncertain parameters and a comparison between experimental and simulation results.

20.5.5 The DEXTER complete model

In order to attain the complete model, the manipulator has been split into subsystems (blocks) according to the modules features. The shoulder, the arm and three wrist links modelled by the LINK module.

Additional modules have been made to model the parallelogram mechanism, the gravity balancing mechanism, the mechanical interactions among the wrist links, the load. Finally an ad-hoc module has been developed for the base link whose equations are simpler since it is the first link in the chain and a prismatic one. For the sake of brevity these last modules will not be described.

By assembling all the blocks according to LEGO rules, the complete manipulator open loop model has been obtained. It consists of more than two hundred simultaneous algebraic and differential equations.

Even if only the forward dynamics model has been described so far, the inverse dynamics model has also been obtained. Indeed, since the same equation stands for both forward and inverse dynamics, obtaining one model from the other is just a matter of interchanging the inputs and outputs of the LEGO modules, and this can be easily done by modifying the module source codes.

20.6 Conclusion

A mathematical model of an industrial redundant manipulator has been presented. The manipulator has a complex mechanical chain. The modelling problem has been tackled by using the LEGO modular simulation code. LEGO has proved to be very suitable for manipulator modelling as it permits the breakdown of the manipulator into blocks and the modelling of each block independently.

So far the model has been used to design and tune an independent joint controller and for the mechanical design of a new manipulator prototype.

At present it is being used to evaluate the improvements which can be obtained using a centralized (multivariable) control scheme of and to find a good trade-off between control computational requirements and performances.

Once the control algorithm is implemented in the controller of the real manipulator, a proper test result analysis will provide a measure of the effectiveness of the manipulator model. Nevertheless we wish to highlight the deep insight into robotics obtained in developing a manipulator dynamic mathematical model.

20.7 References

[1] Luh J.Y.S., Walker W., Paul R.P.C., 'On-line Computational Scheme for Mechanical Manipulators', Robot Motion, MIT Press, 82

[2] Walker M.W., Orin D.E., 'Efficient Dynamic Computer Simulation of Robot Mechanism' Robot Motion, MIT Press, 82

[3] Vukobratovic M. Kirkanski N., 'A Method for Computer-Aided Construction of Analytical Models of Robotic Manipulators', IEEE -CS Int. Conf. ROBOTICS, Atlanta, 84

[4] L. Vecchio, S. Nicosia, F. Nicolo', D. Lentini, 'Automatic Generation of Dynamical Models of Manipulators', Proceedings 10th ISIR, Milan, 80

[5] C. Maffezzoni, G. Magnani, L. Marcocci, 'Computer Aided Modelling of Large Power Plants', Proceeding IFAC Workshop on "Modelling and Control of Electric Power Plants"- Como Italy, Sept. 83, Pergamon Press

[6] L. Marcocci, S. Spelta, 'Computer Aided Modelling of Complex Processes: A Program Packaqe', Proceedings of IMACS Int. Symp. Simulation, in Engineering Sciences, May 83, Elsevier, Amsterdam

[7] Hollerbach J. M., 'A Recursive Lagrangian Formulation of Manipulator Dynamics and a Comparative Study of Dynamics Formulation Complexity', Robot Motion, MIT Press. 1982

[8] Paul R.P., 'Robot Manipulators: Mathematics, Programming, and Control' MIT Press, 1981

[9] Electro-Craft Corp., 'DC Motors, Speed Controls, Servo System', Electro-Craft, Hopkins, MINN. 1980

[10] Luh J.Y.S., 'Conventional Controller Design for Industrial Robots - A Tutorial', IEEE Transactions on Systems, Man and Cybernetics, Vol. SMC-13, 1983

[11] Canudas C., Astrom K.L., Braun K. , 'Adaptive Friction Compensation in DC Motor Drives', Proceedings IEEE International Conference on Robotics and Automation, San Francisco. 1986

[12] Kubo T., Anwar G., Tomizuka M., 'Applications of Nonlinear Friction Compensation to Robot Arm Control', Proceedings IEEE International Conference on Robotics and Automation, San. Francisco. 1986

[13] Magnani G., Bologna P., Rizzi G., Lovati G., 'Modelling and Simulation of an Industrial Robot', Proceedings IEEE International Conference on Robotics and Automation, Philadelphia, 1988.

21. Tread Pattern Tyre Noise: An Approach to Computer Simulation

F. Nava Mambretti

Firestone International Development,

Rome, Italy

21.1 Abstract

This paper presents the basic steps necessary for preparing computer programs able to predict the noise generated by a tyre tread. The geometry of the tread design has to be given as input to the computer that has to be capable to handle such data. The following factors related to tyre dynamic footprint should also be considered: contour of the footprint, pressure distribution, variation of groove depth, possible translations along circumference of each one of the ribs forming the tread.

21.2 Background

The network of highways in all countries has grown during last decades and the uniformity of the road surfaces has been improved: this factor gives rise to a more comfortable ride for the vehicles. Besides, vehicles have been designed to reduce the fuel consumption, result obtained decreasing the resistances to the motion, which implies also a reduction of the total car weight.

While better designs of the automobile engines allowed to reduce vibration and noise produced by them and also their transmission, the tyres, on the contrary, did not show a tendency to have their noise emission decreased because of the structures now in use (like steel tread plies) and of wider tyre treads, with consequent wider footprints.

The situation requires an improvement of the comfort inside the vehicle, thus implying a lower transmission of noise and vibration generated at the contact tyre-road together with a reduction of the noise emission at the source.

21.3 Definition of the problem

The tyre engineers have accepted the challenge to work for achieving these two distinct results: to decrease the propagation, through tyres, suspensions, car body, of the impulses due to the irregularities of road surface, and to reduce the generation of those vibrations and noise that are related to non-uniformities of tread pattern or tyre structure. To decrease the noise propagation means to control the vibration characteristics of tyre sidewalls; to reduce the noise generation requires control of the impact force (vibration input) into contact area, which involves work on rubber compounds and tread design.

This paper will deal only with this second aspect, and it will be even more restrictive, examining only the noise coming, from the tread pattern of a tyre running on a smooth surface (the interaction of the tread with the tyre body will be seen in paragraph 21.6.1, "Tyre modal analysis").

The result to be attained consists in trying to predict the spectrum of the tyre noise emission, and to modify it at the stage in which the 'tyre' is still on the drafting board (today we should say 'on the screen of the video terminal'), without building green tyres, curing them and test on a vehicle (1).

The hypothesis based on vibration (rather than the one based on aerodynamics, - pumping - once having a large following in USA), will be adopted here for explaining the generation of noise at the tyre-road interface; hypothesis widely corroborated by experimental results, especially for radial tyres (2).

In short, some energy is supposed to be dispersed in vibration, and therefore in noise, when a small block of the tread is released from the trailing edge of the footprint and comes back to its undeformed position; see Figure 85. A similar phenomenon, but in the opposite sense (due to rubber compression instead of expansion), can be expected also on the leading edge.

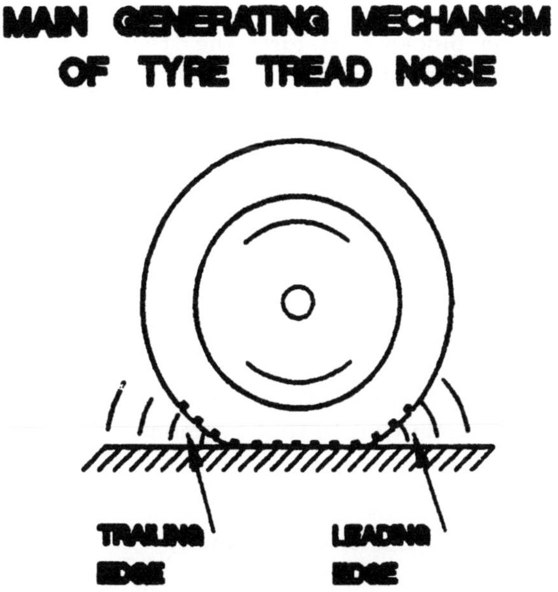

Figure 85. Noise radiation from a rolling tyre

21.4 Description of previous work

About fifty years ago, the tread design became 'noise treated': in order to avoid a 'tonal noise', with an emission concentrated about only one frequency and its harmonics, the unit of design that repeated itself identically in the tread pattern was substituted by a sequence of different units that are called pitches.

Formulas and procedures for determining pitch lengths and for obtaining a pitch sequence effective in reducing tonal noise, have been available for many years. For some of the most interesting works on this subjects see [3], [4]. A patent request is pending for a new method that has been derived by application of electronics techniques for frequency modulation of signals: like the previous ones, the object is to uniformly distribute the acoustic energy over a wide frequency band, each frequency containing the lowest possible energy.

21.4.1 Results of computer introduction

Regarding the use of computer for prediction of noise frequency spectra, some mathematical models have been in use for 15-20 years: they give the lengths of the blocks that form the tread and their sequence along the tire circumference; for this aspect see [5], [6], [7]. A frequency distribution spectrum of noise produced by a given pitch sequence, for given revolutions per second of a tire, can be obtained; because frequencies generated are linearly dependent on speed, if the speed doubles, the frequencies of the peaks of the spectrum will double too.

These methods can be derived by an elementary application of the series of Fourier, in order to pass, from a succession of pulses, succeeding at non-constant intervals, to the spectrum of frequencies generated. One harmonic is examined at a time, and each pitch length is evaluated for its contribution to that harmonic. The application of the basic formula

$$f(t) = \frac{a_0}{2} + \sum_{n=1}^{\infty} (a_n \times \cos nt + b_n \times \sin nt)$$

is not difficult, because the search of the coefficients $a_n$and b_n , is simplified supposing all the pulses being identical in amplitude, and the phase is determined by the position of the point along the circumference.

21.4.2 Operating techniques

First of all, the number of blocks of the tread pattern has to be determined, then the tyre designer decides how many different pitches have to be used (three or more) and calculates their lengths. After an initial sequence has been given to the computer, the program can be run. Once the spectrum of frequencies is obtained, a change can be introduced in the sequence, and the effects on the spectrum will be evaluated. This last part of the process can be automated. When high amplitudes of peaks are found in the spectrum, or crowding of high peaks, the sequence and/or the lengths of the pitches have to be modified, in order to decrease the amplitudes of the impulse corresponding to the frequency related to those peaks.

Today, with computer of significant power, thousands of random sequences of variously selected pitch lengths can be analyzed, which can yield good results in short time.

21.4.3 Limits of traditional approaches

The actual tread design is not considered in these mathematical approaches, or better said, they look at a design made only by transverse slots, perpendicular to the rolling circumference; see Figure 86.

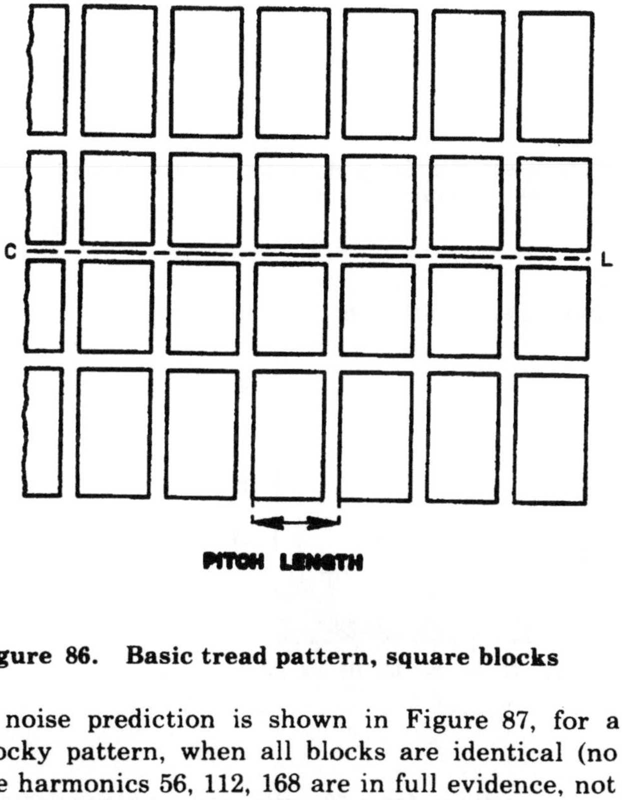

Figure 86. Basic tread pattern, square blocks

A noise prediction is shown in Figure 87, for a 56 pitches of a blocky pattern, when all blocks are identical (no noise sequence): the harmonics 56, 112, 168 are in full evidence, not attenuated. With a good sequence the shape of the graph can be modified according to Figure 88, valid again for a 56 unit design, with three different pitches.

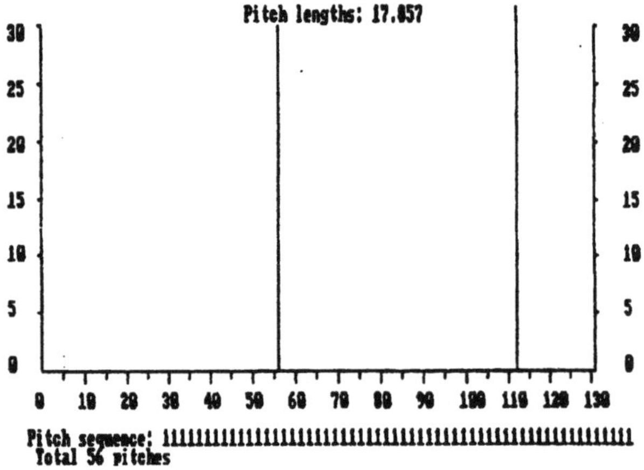

Figure 87. Frequency spectrum for a 56 units, square block design, without noise treatment

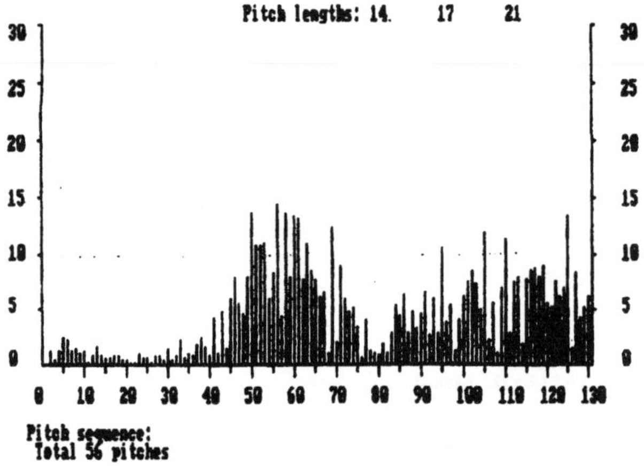

Figure 88. Frequency spectrum for a 56 units, square block design, with noise treatment

More recent models take into consideration also the width and the depth of the slots; that means the signal producing the noise is not assumed to be generated along a contour line, but all over a surface (area of the slot). Then, according the vibrational hypothesis of noise generation, in which bigger the mass of the vibrating block higher should be the impulse resulting, the amplitude of the signal should also be changed, in accordance to the variation of groove depth.

Theoretical prediction of the spectra and experimental results (noise recordings, made in laboratory of a tire rotating on road-wheel, studied with a F.F.T. - fast Fourier transform - analyzer, to find the experimental frequency spectrum), are in good agreement, as far as the location of the main frequencies of noise emission is concerned, but there is little conformity for the amplitudes of the peaks of these frequencies (often peaks that seem highest in the prediction, are not found to be such in experiments).

It is evident the need to find a procedure which would handle the difference in phasing of the various ribs that form the tread, or, even better, considers the actual full pattern that forms the tread design.

21.5 New approaches to noise simulation

Only a preliminary selection of sequences can be obtained with the methods mentioned above, or better, these methods can be used to prepare a library of sequences, optimized and stored according different number of slots, according a given number of different pitches (from 3 up to 7), according a given ratio between longest and shortest pitch (ratio that should vary from truck tyres, to commercial van, mud and snow, or passenger car tyres); how this library can be utilized will be seen later in this paragraph.

21.5.1 Need of the actual tread design

The base for obtaining an effective noise prediction is the actual tread design [1]. This result can be achieved today in a practical way, due to the availability of computers with high capacity (and speed) of calculation, and of interactive graphic terminals of high quality, with proper software for geometrical design: the so called CAD systems. All this, in order to allow the tyre designer to draw the basic unit of the tread (Figure 89) and to introduce quickly all

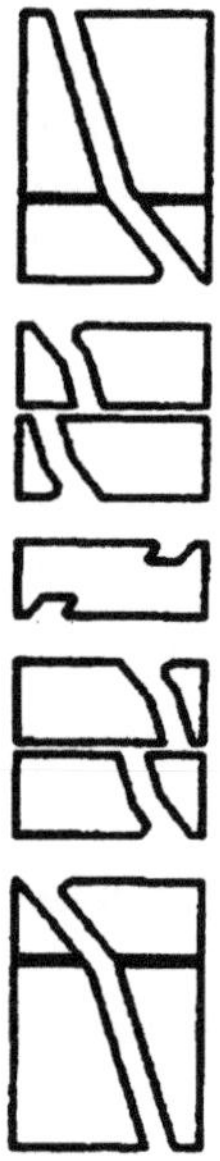

Figure 89. Basic pitch of an actual tread design

the modifications that, during the analysis, seem to become useful (Figure 90). Even a program not based on an interactive terminal can achieve the same objective, but more slowly and laboriously (giving numerical coordinates of all vertices of the design).

The spectrum of frequencies, obtained when the calculations are executed, has to be evaluated, with the basic criteria seen before: avoid sharp peaks, try to have a 'white noise' profile.

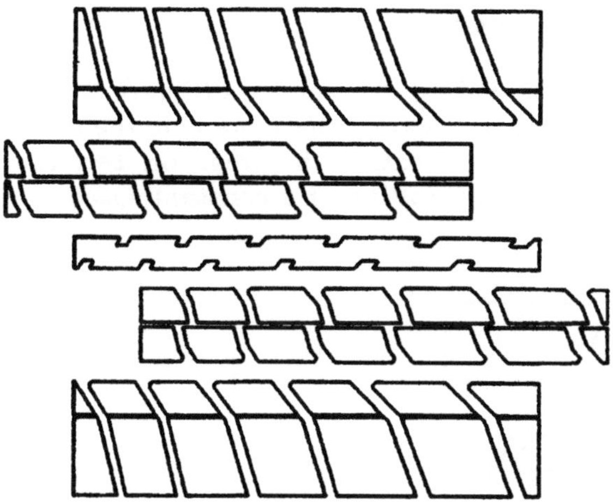

Figure 90. Tread pattern with offsets

It has to be noted, in relation to the work [8] and similar studies carried on in Germany, that the tyre designer cannot take his decisions about pitch sequence modifications, only listening at a noise, synthesized on the basis of a prediction and reproduced through a loudspeaker.

This technique can be useful for a better evaluation of the phenomenon, but it is only the analysis of the frequency spectra that can indicate where and how to operate. A mere listening of the noise can allow to make a choice among various noises, but does not tell nothing about how to do in order to improve the quality of noise emission.

21.5.2 Need of dynamic footprint

The real footprint is not delineated by straight lines (Figure 91), but rather by curved lines; the program should therefore give the option either to use circular arcs for delineating the tread or, better, to input the true footprint of the tyre, rolling at a constant speed relevant for noise emission.

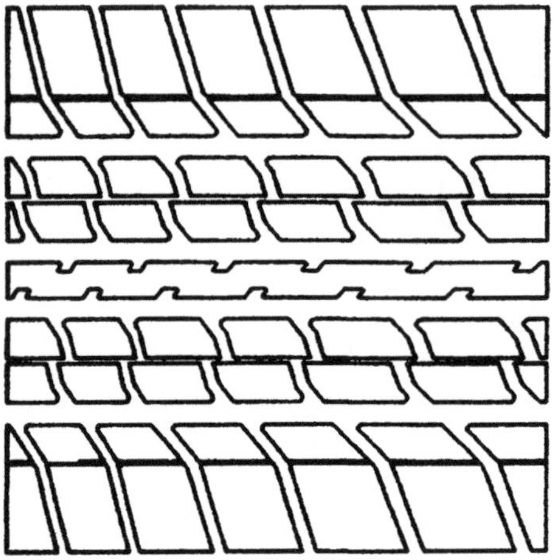

Figure 91. Sequence of different pitches

Today it is possible to have the exact contour of the footprint for a tyre rolling under a car. The footprint contour can be 'digitized' and reconstructed by the computer that will lay it upon the tread design and will use it for scanning the tread (Figure 92). But this contour can change with the speed, with the weight of the car, with the tyre inflation pressure, and this will modify the predicted spectrum.

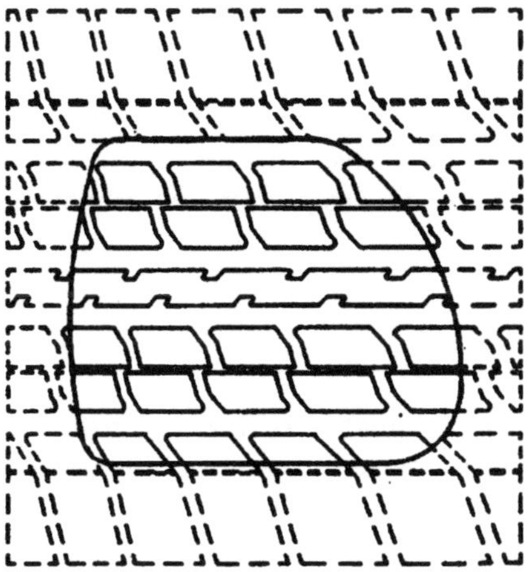

Figure 92. Footprint contour, over impressed on tread design

All this can lead to calculations finalized to a specific car application (with difference, in noise emission, between front and rear tyre position): this is not bad, because the tyre has to be developed looking to the final use, which means the vehicle ride. Obviously, for a tyre that has to be fitted on cars of different characteristics, a compromise has to be found among various possible functions of the tyre tread. The ideal for the tyre engineer is to design a tread, that keeps to a minimum the sensitivity of the tread pattern to footprint changes.

21.5.3 Additional factors to be considered

An uneven distribution of pressures in the footprint, measured at a given speed, is another factor that has to be considered: it can have a significant effect on the shape of the noise spectrum.

Also variation of the groove depth can be taken into account, with corrective coefficients: lower the depth of the groove, smaller the movement of the edge of the block of rubber. When the depth becomes one half, the energy emitted as vibration is reduced approximately to one quarter.

21.6 Future trends for a complete noise modeling

Most of the noise testing carried out in a laboratory, with tyres rolling on road-wheel, can be eliminated with the method described above, because it is easy to know in this case the footprint and all data needed for obtaining at computer the frequency distribution of a given tyre design.

21.6.1 Tyre modal analysis

The effects of resonances due to the geometry of tyre volumes and to tyre structure (related to physical-elastic properties of materials used for body and belt, and of rubber compounds) have to be taken into account in order to obtain a full simulation of the actual noise generated by a tyre.

Nowadays this can be done through the 'modal' analysis of a tyre. With an F.F.T.- dual channel analyzer - the response to a forced vibration, induced by an external exciter in various points of a tyre, can be measured: the natural frequencies of resonance (the 'modes') can be obtained and using special computer software the physical meaning of them can also be evaluated. This structure-vibrational model of the tyre can be compared with the one obtained for the tread emission, and, for a given speed, the ranges of frequencies that can create problems, can be deduced.

Also influence of road surface texture has to be considered [9]: but this aspect of the question is outside the scope of this paper.

21.6.2 Transmissibility inside vehicle

The next step is to consider not the noise as it is emitted by the tyre, but as it is really perceived inside the vehicle. Here we just mention a couple of the approaches that can be used for solving the problem.

In a given vehicle, the 'transfer functions' of car body and suspensions can be determined by experiments, studying how the impulses, generated by four shakers applied to tyre axles, pass inside, and in consequence, how the frequencies emitted by the tyres can be filtered or amplified, on their way into the vehicle.

The effects of tyre noise inside the vehicle can be studied also measuring the transfer functions in a car, equipped with plain tread tyres, rolling on roadwheels. Once the spectrum of the noise generated, in free field conditions, from the tread under study is known, the effects of combining this spectrum with the previous transfer functions can be evaluated.

What has been said until now is related to the control of noise emission of one individual tyre: but the tyres emitting noise, are four at a time, and they are not in phase; in a vehicle there are other rotating parts (engine, transmission) producing noise: therefore effects of phasing like modulation and similar audible phenomena can still be possible.

21.7 Summary

An information, not complete, but significant and useful, about the noise spectrum related to a given tread design can already be obtained: it can be verified which, of a series of proposed designs can be less 'noisy', and, for a given design, which, of a group of selected pitch sequences, will result in a quieter tyre ride.

In the near future the simulation of the effect, inside a given car, of a given type of tyre, with a given design and noise treatment, will become possible, without building the tyre and with elimination of many testing operations.

A tyre optimized for a given car might not be the best on other cars. A specific noise treatment of a design, valid for a given tyre size (e.g. 175R13), can be less acceptable for another size (e.g. 185/70R13): as a matter of fact this type of work will always require careful studies and checks to get a better product. The search for a compromise solution, valid for more than one car, will remain a task that now, and in the future, requires skill, experience and sensitivity of the tire designer.

21.8 Acknowledgements

The writer wish to thank The Firestone Tyre and Rubber for permission to publish this study.

21.9 References

[1] F. Nava Mambretti - Tyre noise simulation at computer - Advances in control and refinement of vehicle noise - IMECHE - London - March 1988. This work contains the basics of the theory exposed here.

[2] P.E. Waters - Characteristics and mechanism of tyre noise. Proceedings of International tire noise conference, Stockolm, 1979.

[3] F. Campos -Y. Herbelleau - Reduction of the travel noise of tires U.S. patent 4 598 748 July 8, 86.

[4] K. Ishikawa -T. Kogure - Pneumatic tire. Japan patent 1980-8904.

[5] J.H. Varterasian - Quieting noise mathematically - Search. G.M. Res. Lab. Sept/Oct. 1967.

[6] P.R. Willett - Tire tread pattern sound generation - Tire science and technology - Nov. 75.

[7] W.J. Vorih - Designing quiet tread spacings for tires. 99th rubber division meeting on AM. Chem. Society - 1971.

[8] D.H.Parker - D.G.Caruso - D.B.Thrasher - R.J.Blinn - Method of simulating tire tread noise and apparatus- U.S. patent # 4 727 501

[9] J.C. Walker - D.J.Evans - Improvement of road excited internal vehicle noise. Proceedings of 16th ISATA Symposium - May 87.

22. Some Examples on the Utilization of Mathematical Tools for the Design Application in LABEN Experience

G. Alberti, M. Biserni, A. Colombo

LABEN

SS. Padana Superiore 290

20090 Vimodrone (MI)

Italy

22.1 Introduction

Since its foundation Laben has been engaged in the design and development of electronic systems. One major branch of application of these systems is related to the implementation of equipments for space scientific experiments and payloads.
These equipments typically embed: electronics, thermo-structural parts, critical and special items that in turn have to perform complex scientific experiments tasks.

Due to the always changing application nature of the equipments, careful simulation and validation phases have to be performed even at the very beginning of the research and development studies and in parallel with them.

The growing availability of "turn-key" CAD/CAE programs can help the design engineer in its development effort, but, very often, these "standard" packages cannot yet cover all the specific design peculiarity or does not provide complete access and flexibility for the user to face dedicated problems.
To cover these gaps, therefore, suitable mathematical alghoritms and tools need to be "in house" developed and proven.

This paper want to show some of these internally Laben developed tools with the aim to provide their schematic overview (for a complete explanation the reader is addressed to ref. [2-4-5]).

Three different (but equally meaningful) areas of implementation are presented:

- mechanical analysis and simulation
- thermal analysis
- special x-ray imaging application

It must be noted that on the above topics the internal activities of refining, finalization and verification are continuously running, particularly from the standpoint of the merging of these specific tools with the standard and general purpose CAE/CAD system that we are at the same time utilizing for the design.

22.2 Mechanical Simulations Analysis: Needs and Tools

A recurring problem in the aerospace field is to characterize dynamically any structure when it is still in a planning phase or even during the feasibility study.
Of course well known and qualified software packages can solve this kind of problems (NASTRAN, ADINA, etc ...), but some drawback exist:

- They are not always able to solve particular problems
- Their use needs a long training time
- They are expensive

Thus, in some cases the development of in-house-made software could be convenient and cheaper.

22.2.1 Frequency Response Analysis Program (FRAP)

The rationale of the Process starts from a Finite Element Modeling (F.E.M.) of the structure under study and follows the annexed flow chart (Figure 93).

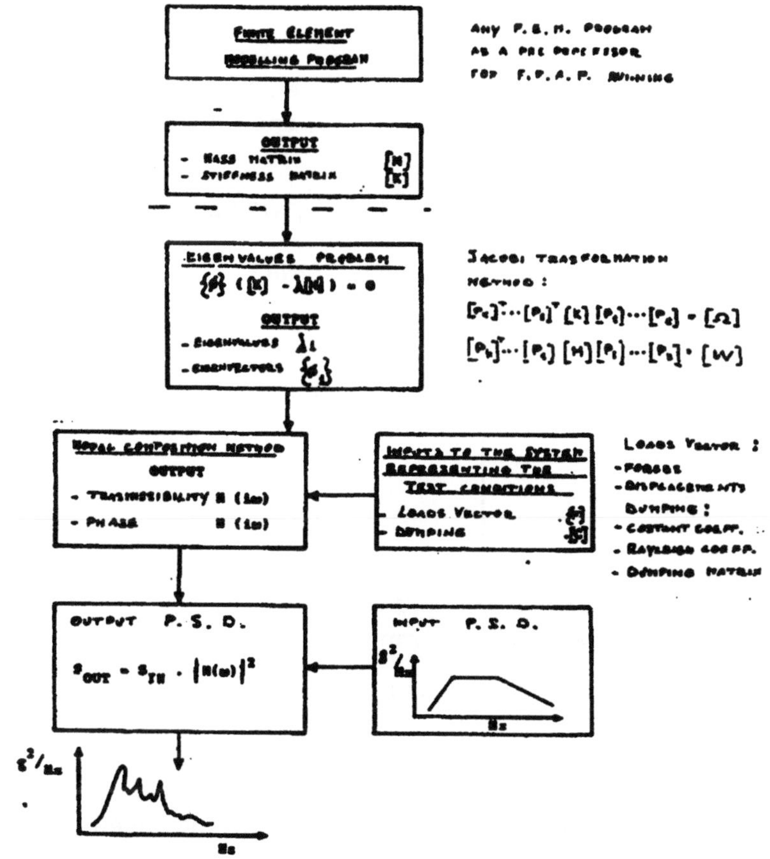

Figure 93. Frequency Response Analysis Program Flow-Chart

The major aim of the FRAP code is to simulate the behavior of a structure when submitted to one of the most critical task foreseen for space equipment acceptance: the so called "random vibration test". Typical test conditions involve dynamic random excitation over a frequency range from few Hertz up to 2000 Hertz for few minutes duration and up to about 20 g RMS accelerations.

The calculus of the matrices $[M]$ and $[K]$ can be done by any F.E.M. solver and it represents the essential pre-processing for the following dynamic analysis.

The eigenvalues problem $([K]-\lambda[M])\ \{\varphi\}=0$

is solved by the "JACOBI" transformation method based on the orthogonality property of the eigenvectors ref. [1]. This algorithm diagonalizes contemporarly the matrix $[M]$ and $[K]$ by means of a series of matrix products:

$$[P_k]^T\,[K]\,[P_k] \quad \text{and} \quad [P_k]^T\,[M]\,[P_k]$$

the rotation matrix $[P_k]$ has the following form:

$$[P_k] = \begin{bmatrix} 1 & & & & & \\ & \cdot & & & & \\ & & 1 & \alpha & & \\ & & \gamma & 1 & & \\ & & & & \cdot & \\ & & & & & 1 \end{bmatrix} \begin{matrix} \\ \\ i \\ j \\ \\ \\ \end{matrix}$$

(columns i, j)

The parameters α and γ are solved in order to zero the related out of diagonal elements of the matrices $[M]$ and $[K]$.
The final results are the diagonal forms:

$$[\Omega] = [P_k]^T \ldots [P_1]^T\,[K]\,[P_1] \ldots [P_k]$$

$$[W] = [P_k]^T \ldots [P_1]^T\,[M]\,[P_1] \ldots [P_k]$$

From the above equations we can derive both the eigenvectors matrix and the eigenvalues vector ,that is:

$$[\varphi] = [P_1]\cdot[P_2] \ldots [P_k] \;\; ; \quad \lambda_i = \Omega_i \,/\, W_i$$

Now we can rewrite the dynamical system as a series of uncoupled equations having the form ref. [2]:

$$[W]\,\{\ddot{x}\} + [D]\,\{\dot{x}\} + [\Omega]\,\{x\} = [\varphi]^T\{R\}$$

Of course also the dumper matrix must be diagonal;for this reason three dumping cases are foreseen by the FRAP code:

- Constant dumping coefficient
- Rayleigh dumping coefficient
- Dumping matrix

Once the user has defined one of the above dumping alternatives the program is also able to handle the relevant algorithms for computations (not discussed here for sake of brevity).

Now, assuming for the node displacement X a sinusoidal form we can write the dynamical system as:

$$- w^2 \, W_i \, \bar{x}_i + i \, w \, D_i \, \bar{x}_i + \Omega_i \, \bar{x}_i = \lfloor \varphi_i \rfloor \, \{F_0\}$$

Using the modal composition method

$$\Delta_i = \sum_{j=1}^{n} \varphi_j^{(i)} \, x_j$$

we can estimate the node displacements as

$$\{\bar{\Delta}\} = \sum_i \frac{\{\varphi_i\} \lfloor F_0 \rfloor \{\varphi_i\}}{\Omega_i \left[1 - \left(\frac{w}{w_i} \right)^2 + 2 \, i \, \xi_i \frac{w}{w_i} \right]}$$

This relation is still true even if the input stimulus imposed to the structure is a displacement instead of a force. In this case the loads-vector assumes the form:

$$\{L\} = Y_0 \, \{(K_v + i \, w \, C_v)\}$$

and the equation

$$\left\{ \frac{\bar{\Delta}_0}{Y_0} \right\} = \sum_i \frac{\{\varphi_i\} \lfloor (K_v + i \, w \, C_v) \rfloor \{\varphi_i\}}{\Omega_i \left[1 - \left(\frac{w}{w_i} \right)^2 + 2 \, i \, \xi_i \frac{w}{w_i} \right]} \equiv \mathrm{H}(i \, w)$$

gives the trasmissibility function directly.

Now knowing the Power Spectral Density imposed in input to the structure we can obtain the Power Spectral Density (PSD) output to the considered nodes from the relation:

$$(\text{P.S.D.})_{OUT} = |H(i\,w)|^2 \cdot (\text{P.S.D.})_{IN}$$

As an example we have modelled an electronic board by means of a FEM software. FRAP calculates transmissibility, phase shift, PSD for each selected node. Knowing the input PSD we were able to derive the response of the structure in the point (Figure 94).
The analyzed item was tested with the same input conditions and the result is depicted in Figure 95.
We have to note the good agreement between the analysis and the test, fact that makes us confident with the mathematical method and with its software implementation.

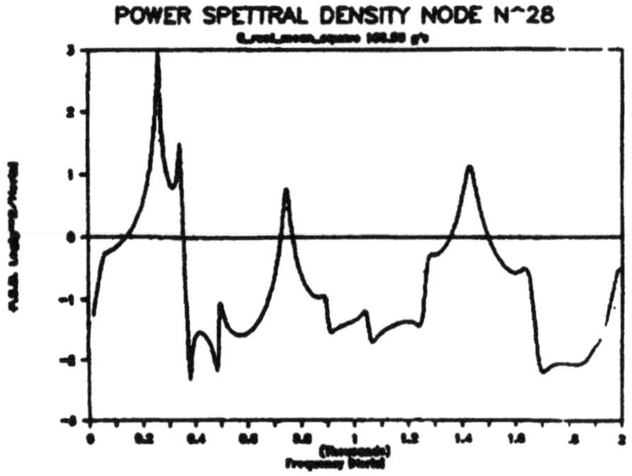

Figure 94. Simulated Response of a Structure Obtained by FRAP

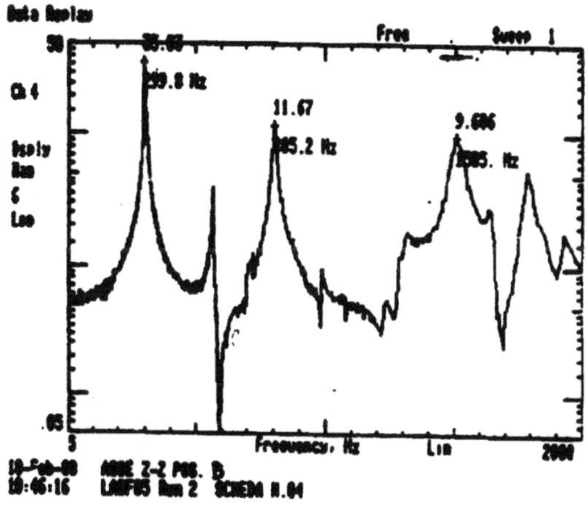

Figure 95. Measured Response of the Structure Previously Simulated

22.3 Thermal Analysis Needs and Tools

As already pointed out in 22.2, "Mechanical Simulations Analysis: Needs and Tools," the nature of the tasks and operating environment of spacecrafts dictate that they must be high integrity sistems. Also from the standpoint of the thermal problems, a careful analysis and simulation of the devices to be flown is therefore needed in order to prevent any potential source of failures or degradation.

For the above tasks, the lumped parameter method is currently used. It is well established and widely accepted by thermal analysts. The system under analysis is discretized into cells.
Each cell is associated with a node and with its thermal capacitance and temperature. Thermal couplings among the nodes are calculated on the basis of the geometrical and physical characteristics of the cells (this normally requires a great burden of work, especially for irregular grids of nodes and for the computation of radiative exchange factors). The main existing numerical codes based on the lumped parameter method used for space applications are the well-known SINDA (Systems Improved Numerical Differencing Analyzer) and ESATAN (European Space Agency Thermal Analysis Network) which is basically a replacement for SINDA offering, in addition, an interactive context editor and a sub-modeling facility. However,

these programs require a high level of user training and familiarization and need quite large computer resources.
Therefore LABEN has developed an in-house software based on the lumped parameter method with reduced capabilities with respect to SINDA-ESATAN but with a simpler user interface. The program, written in PASCAL and named LISTA (LABEN Internal System Thermal Analyzer), runs on an IBM PC and will be described hereunder.

22.3.1 LABEN Internal System Thermal Analyzer (LISTA)

Figure 96 shows the rationale of the process.
Spatial and time discretization are introduced by LISTA for the thermal analysis of a system. The extent of the discretization depends from the required precision.

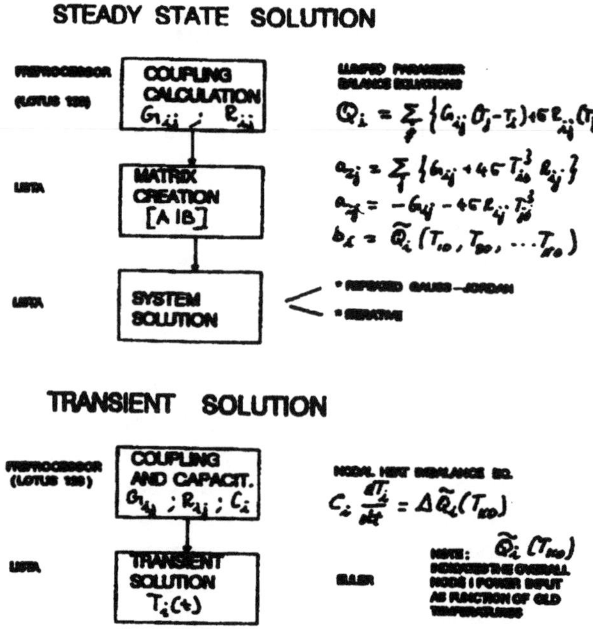

Figure 96. Thermal Analysis Steps Involving LISTA Program

The equations of the lumped parameter approximation can now be stated in the following form which includes implicitly the boundary conditions:

$$C_i \frac{dT_i}{dT} = Q_i + \sum_{j=1}^{N} \{G_{ij}(T_j - T_i) + \sigma R_{ij}(T_j^4 - T_i^4)\}$$

where

T_k = temperature of the k-th node,
C_i = thermal capacitance of the i-th cell,
Q_i = heat dissipated within the i-th cell,
G_{ij} = conductance (or linear coupling) between node i and node j,
R_{ij} = radiative exchange factor between i-th and j-th nodes,
σ = Stefan-Boltzmann constant.

The thermal modeling problem has been therefore reduced to the solution of a set of N ordinary (non-linear) differential equations in the N unknown T_k.

The program:

Input data for LISTA consist in the specification of the following parameters:

- number of total and of boundary nodes in the model,
- linear (G_{ij}) and radiative (R_{ij}) couplings,
- thermal capacitances (C_i), in the case of transient analysis,
- power inputs (Q_i),
- initial temperature of all nodes.

Data can be input either interactively via the keyboard or by means of a disk-file. The program is interactive and the user has to choice the type of analysis (steady-state or transient) and to specify the output media (screen, disk-file or printer). Calculated output include nodal steady-state temperature and heat exchanges or time evolution for each nodal temperature, depending on the selected analysis type.

Steady-state solution:

putting $\dot{T}_i = 0$ in the above equations, one obtains a system of N non-linear equations corresponding to the steady-state condition. The solution of the system requires several iterations due to the presence of the radiative non-linear terms. These terms are factorized using a first order approximation:

$$T_k^4 = T_{k0}^4 + 4\, T_{k0}^3\, \delta T = 4\, T_{k0}^3\, T_k - 3\, T_{k0}^4$$

where T_k and T_{k0} represent the temperature of the k-th node at a certain iteration and at the previous one. Substituting this factorization into the radiative terms of the equations leads to a new linear system which, written to matrix notation, reads:

$$A\, T = B$$

A is a square matrix with NU rows and NU columns, where NU is the number of unknowns (i.e. $N - NB$ with $NB =$ number of boundary nodes); T is the unknown vector and B is a known vector. The elements of A and B, calculated directly by LISTA at each step, are:

$$a_{ii} = \sum_{j=1}^{N} (G_{ij} + 4\, \sigma\, T_{i0}^3\, R_{ij}) \qquad i = 1, 2, \dots \text{NU}$$

$$a_{ij} = -G_{ij} - 4\, \sigma\, R_{ij}\, T_{j0}^3 \qquad j \neq i$$

$$b_i = Q_i + \sum_{j=1}^{N} 3\, \sigma\, (T_{i0}^4 - T_{j0}^4)\, R_{ij} + \sum_{j=\text{NU}+1}^{N} [(G_{ij} + 4\, \sigma\, R_{ij}\, T_j^3)\, T_j]$$

The linear system is solved using the direct Gauss-Jordan elimination method operated on the augmented matrix $(A \mid B)$. It is important to note that, according to their definition, all the diagonal elements of A are different from zero, which allows to proceed safely with the normalization of the rows.

Transient solution:

The transient solution is based upon the calculation of the nodal imbalance of power. A temperature change can be calculated if the imbalance remains constant for a small interval of time dt:

$$C_i \cdot \frac{T_i(t + dt) - T_i(t)}{dt} = \Delta Q_i$$

LISTA determines the time step *dt* to be used in the integration on the analogous electrical *RC* constant. The *RC* factor for each node is calculated according to the following definition and the minimum value determined is selected. The user can however modify the calculated *dt* by means of a multiplicative factor in order to achieve a better accuracy or to speed up the solution. The calculation stops when the time reaches a specified end value.

$$R\,C = \frac{C_i}{\sum_{j=1}^{N}\left[G_{ij} + \sigma\, R_{ij}(T_j^2 + T_i^2)\,(T_j + T_i)\right]}$$

The LISTA code is presently under utilization for a number of internal studies and preliminary test results show good agreement between expected nodal temperatures and experimental measurements.

22.4 Special Cases of Study and Algorithms

In space applications it could arise that processing methods used successfully on ground to obtain the desired results, are not viable in the space because of the low available volume on the satellite and the severe environmental conditions.
This particularly holds for scientific payloads that usually requires dedicated and special signal/data processing solutions.

For example the High Pressure Gas Scintillation Proportional Counter under study for S.A.X. satellite is a position sensitive detector for *x*-ray astronomy; this instrument needs a mathematical algorithm able to detect the *x*-ray scintillation point in order to evaluate correctly the incoming energy.
This process normally is performed by two different techniques:

- Anger Camera reconstruction
- Maximum a Posteriori calculation

These techniques however requires computational means that even available on ground laboratories, are not available on board spacecraft. For these reasons, in order to make the process viable, a new technique (the so called "look-up table" method) has been developed. The advantages of such approach are the following:

- Full Hardware implementation
- Low processing time
- No work for the instrument Microcomputer
- Relatively low power consumption of associated HW

This approach will be detailed in the following.

22.4.1 HP-GSPC Reconstruction Algorithm

The basic concept of the technique is:

- Divide the instrument scintillation plane in many areolas (pixels)
- Map in a table each areola by means of a set of special parameter values identifying the areola
- Reconstruct ,for each event, the scintillation areola comparing the relevant parameter set coming from the PMs signals with those mapped
- Correct the event energy multiplying the Sum-signal by the energy correction coefficient associated to the scintillation areola

It must be noted that the last step is due to the fact that the sum signal (proportional to the incoming energy) depends also from the entry point over the detector. Taking into account the particular symmetry in the phototubes distribution, which is shown in Figure 97, we can reduce the mapping to 1/12 of the sensitive surface.

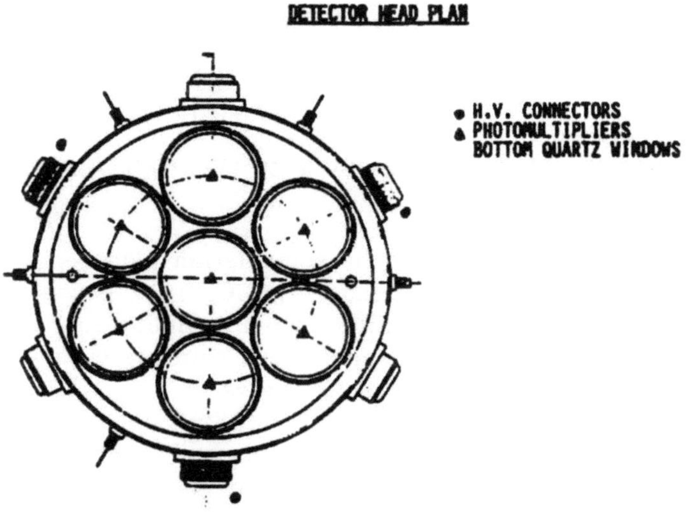

Figure 97. Detector Lay-Out

The scintillation areola is identified on the basis of the intensity of the signals supplied by the 7 PMs and processed by the counting chains. Obviously such signal intensities depend on the incident x-ray energy which is a priori unknown. Therefore the parameters allowing the areola identification must be:

a) Independent from the incident x-ray energy
b) Dependent from the coordinates of the scintillation point

If we don't care of the statistical effect, we have:

S_j = Signal coming from the j-th PM
N_{phj} = Number of photons picked up by the j-th PM
N_{phg} = Number of photons of the scintillation process
$\Omega_i(x;y)$ = Solid angle under which the scintillation point see the j -th PM

$$S_j \div N_{phj} \div N_{phg} \cdot \Omega_i(x;y)$$

and

$$S_{TOT} \div N_{phT} \div N_{phg} \cdot \Omega_{TOT}(x;y)$$

Therefore the coefficient $C_j(x\,;y)$ defined by the equation:

$$C_j(x\,;y) = \frac{S_j}{S_{TOT}} = \frac{\Omega_i(x\,;y)}{\Omega_{TOT}(x\,;y)}$$

is directly obtainable by the detecting system and meets both a) and b) requirements.

Since at every point P(x;y) of the scintillation plane is associated one and only one set of coefficients $C_1, C_2, C_3, \ldots C_7$, it could be possible to come back to the scintillation point finding that areola for which all the seven tabulated coefficient coincide with the seven coefficients achieved by the counting chains.

Now if we tabulate for each pixel the upper and lower limit value of each coefficient, we come to define seven bidimensional intervals relevant to the same areola in the coefficient space.
It will be possible to trace back the scintillation areola singling out that pixel whose intervals contain the relevant coefficients.

Due to statistical effects on the PMs signals, it is quite improbable to meet all the seven coefficients in a single areola specially for low energy events. For this reason and in order to simplify the research and to minimize the mapping, only a reduced set of PM's is used by the reconstruction technique.
Moreover only the PM's where the signals are expected to be higher (thus reducing statistical effects) are selected for the set.

The simulation program takes into account the statistical fluctuation in the PMs signals in the way depicted in Figure 98. As we can see all the optical effects at the PMs windows have been neglected because they don't change the process characteristics.

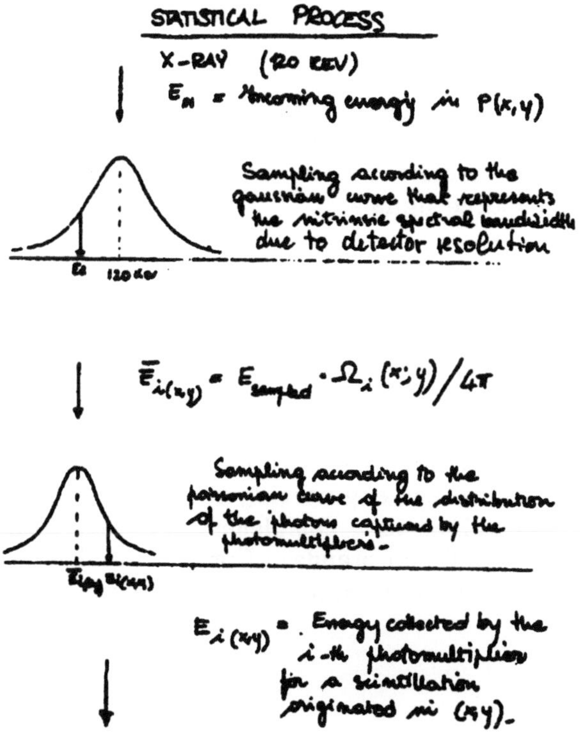

Figure 98. Statistical Effects Taken into Account

The solid angles calculations were performed writing the integral expression

$$\Omega_p = \iint_A \cos\gamma \frac{dA}{R^2}$$

as an expansion in binomial series followed by term by term integration which leads to a series converging for all the values of the coordinates.

More specifically the solution of above solid angle computation is given by the ZUMWALT formula given in ref. [3]. This quite complex equation can be easily solved by the computer and it offer an accuracy more than sufficient for our purposes.
As a test, we have simulated the detection of 2000 events due to 4 calibrations sources located in 3 symmetric positions over the

detector together with a cosmic uniform monochromatic external source.

As a result of this simulation session, the algorithm proved to be able to reconstruct the incoming event position and total energy spectra, being also able to discriminate calibration source signals from cosmic simulated source signals.

22.5 Conclusions

The described mathematical algorithms and tools, aimed to solve research and development specific design problems oriented to the implementation of space equipments, are, according to the authors point of view, good examples of how applied mathematics disciplines can well cope with industrial and engineering needs.
These types of tools, combined in a complementary way, with the growing availability of significant standard Computer Aided Design packages could better support the engineer for the enhancement of its design capabilities.

22.6 References

[1] F. Cesari "Metodi di calcolo nella dinamica delle strutture" Pitagora-Bologna.

[2] Laben Technical note T1 3538, march 1988.

[3] A.H. Jaffey. Review of Scie. Instr. 25. No. 4, pag 349

[4] Laben Technical note "LISTA Handbook", sep. 1987

[5] Laben Technical note TI 2924, march 1987.

23. The Experience at TECNARS of Using Mathematical Modelling

V. Recchi

TECNARS

23.1 Abstract

The experience of using mathematical modelling in solving design problems in TECNARS, a research and development centre for technological innovation in industry, is presented.
Applications are in the field of combustion, heat transfer, fluid flow, design of heat exchangers and units for air conditioning, heat pumps, etc..
The validation is usually made in the experimental laboratory of TECNARS. Problems encountered in mathematical modelling are briefly discussed.

23.2 Introduction

This talk deals with the experience of using mathematical modeling in TECNARS - Tecnologie Avanzate Ricerca e Sviluppo SpA , a private research and development center for technological innovation in industry.

TECNARS is a recently (1987) established company, located in the industrial area of Bari; notwithstanding its young ness, its experience takes origin from the Research and Development Division of an industrial manufacturing company, established in 1974, that was producing boilers, burners, and almost all the types of units for the air conditioning sector.

That Research and Development Division is now organized in the autonomous Company "TECNARS" belonging to the Industrial Group "Acqua" representing its Research and Development Center for Technological Innovations; at the same time TECNARS addresses firms which intend to carry out research and development programs on product/processes, supporting their internal resources with its own project teams, advanced instrumentation and laboratories.

The total technical staff is composed by 22 employees, 8 of which are university graduated engineers; there is an experimental laboratory and a chemical analysis laboratory.

TECNARS operates in the following fields:

- ENERGY, with particular reference to residential and industrial heating and air conditioning;

- ECOLOGY, ranging from measurement and analysis of pollutant emissions to treatment of gaseous effluents;

- COMBUSTION, ranging from characterization of fuels, flames and emissions to design of burners and heat generators;

- TECHNICAL SERVICES, as tests of appliances, special measurements and feasibility studies.

Almost all the research and development activities in TECNARS are involving, even at different level, the use of mathematical models. I want to emphasize that mathematical modeling is far such an organization, one of the means to carry out the applied research on design problems that the industry our customer are committing to TECNARS; from this point of view we consider it as a "tool".

The following give a short overview of the application of mathematical modeling used or developed within the different fields of activity at TECNARS.

23.3 Steady-State Computer Design Model for Air to Air Heat Pumps

Project leader: V. Bruno

This model is a Fortran IV computer program to predict the steady-state performance of vapor compression, electrically driven, air to air heat pumps in both cooling and heating operations.

It is an analytical design tool for use by heat pump studies manufactures, consulting, enginners, research institutions, and universities in directed toward the improvement of heat pump efficiency.

The philosophy of the model development has been to base the program on underlying physical principles and generalized correlations to greatest extent possible, in order to avoid the limitations of empirical correlations derived from manufacturers' literature.

The basis of this model formulation is the logic that links the analytical models of heat pumps components together in a format requiring an iterative solution of refrigerant pressure, enthalpy and mass balance (Figure 99 on page 312).

The program is organized into a modular format so that researchers will be able to readly adapt it to their specific needs.

The model allows the user to specify:

- System operating conditions
- Compressor characteristics
- Refrigerant flow control device
- Fin and tube heat exchanger parameters
- Fans and indoor duct characteristics
- Type of refrigerant

It will compute:

- System capacity and COP or EER
- Compressor and fan motor power consumptions
- Coil outlet air dry and wet bulb temperatures
- A summary of the refrigerant-side states through the cycle

- Overall compressor efficiencies and heat exchanger effectiveness.

This model was developed at OAK RIDGE NATIONAL LABORATORY.

TECNARS acquired this program because of its tasks carried out in research and development on advanced heat pump systems within the International Energy Agency (IEA).

The program, initially developed for running on IBM 3033 computer (32 BIT), has been adapted by TECNARS to run on a HP 1000 computer and has been used for designing both heat pumps and air conditioning units. At the present a personal computer version is available.

23.4 Steady-State Computer Design Model for Water Heater Heat Pumps

Project leader: V. Bruno

This computer design model is a modification of the above described model. It compute the steady-state performance of vapor compression electrically driven water to air heat pumps.

The modifications consists, essentially, in adding the new subroutines:

- a water to refrigerant condenser model
- a circulating water pump model.

The water refrigerant condenser model includes some particular types of both coaxial tube-in-tube and a shell-and-tube designs.

23.5 Advanced Blue Flame Boiler Computer Model

Project leader: P. Giannattasio

In 1986 TECNARS (at that time otb partecipazioni R&D) developed a computer code for the numerical simulation of the flow and the heat transfer in the "Bluflam" boiler, an advanced, high efficiency, natural gas and oil fueled hydronic system developed by OTB in collaboration with "Battelle" Research Centers of Geneva and Columbus, Ohio.

Bluflam boiler consists roughly of a high intensity burner, where flame is confined in a very small volume, and a cylindrical chamber for the radiant heat transfer which is series connected with convective type channels spirally wound round the chamber itself (see the physical model in Figure 100 on page 313).

The cylindrical chamber, the spiral channels and the discharge collectors had already been treated distinctly and mathematical models together with the pertinent computer codes had been developed by the "Battelle" Research Center of Geneve (spiral channels and discharge collectors) and by the University of Naples (radiant heat transfer in the cylindrical chamber).

The former programs compute the unidimensional flow field and the convective heat transfer in the channels taking into account the hydrodynamic and thermal entry lengths, the changes in the cross sectional area, the channel bending and the mass entrance at discrete points along the spiral channel.

The model for the cylindrical chamber utilizes a flux method (Roesler) to compute the radiant heat transfer in the axial and radial directions through the numerical solution of a set of 2nd order ordinary differential equations.

TECNARS contributed to this research program by revising modifying and linking the different codes in a single computer program capable to compute the pressure drop, the temperature distribution and the heat transfer parameters in the whole boiler by iterating on the different parts of the thermal system until the global mass and energy balances are satisfied.

The numerical model validation was then performed comparing the numerical output with the experimental results from several tests carried out in our laboratories on low capacity Bluflam boilers (BF0). Figure 101 on page 314 and Figure 102 on page 315 show a comparison between some numerical and experimental data of greatest interest.

This validated program was finally employed for the design optimization of a new series of high capacity Bluflam boilers (BF II). The tests carried out on the prototypes of the new boilers confirmed the performances predicted by the numerical model with good approximation.

23.6 Water-to- Refrigerant Evaporators Model

Project leader: P. Giannattasio, V. Bruno

The computer codes available on the market for the design of evaporators and condensers of vapor-compression heat pumps and chillers often provide reasonable results, but they rarely can be employed for optimization studies due to the roughness of the physical models and/or to the restricted range of applicability.

In order to provide reliable, flexible and accurate calculation tools for the design optimization of water-to-refrigerant tube-in-tube and shell-and-tube evaporators, TECNARS developed some calculation models which were validated by means of proper experimental tests.

The tube-in-tube evaporators computer code proceeds "step-by-step" along the axis of the heat exchanger, computing in each node the thermal and flow parameters on refrigerant side and water side for parallel-flow or counter-flow arrangements and in the case of evaporation occurring inside or outside the tubes.

Several correlations from literature depending on the flow regime were employed and very accurate results were obtained. With reference to one of the most sensitive parameter to the model accuracy, Figure 103 on page 316 shows a comparison between numerical and experimental temperatures of the refrigerant at the evaporator outlet.

The development of a calculation model for shell-and-tube evaporators was more complex than the tube-in-tube one, due to the wide range of geometrical arrangements which may occur and to the high complexity of the flow inside the shell.

For these reasons, preliminary tests had to be carried out on water-to-water heat exchangers in order to validate some theoretical model for the- calculation of the shell side hear transfer coefficient and pressure drop.

The computer code proceeds "step-by-step" along the directions of the refrigerant and water flow streams.

The program takes into account that the corresponding points inside and outside the tubes have different abscissae in the reference frames fixed to the two flow streams, due to the presence of cross-flow regions.

Furthermore the code considers that in the same cross section of the evaporator the refrigerant may be evaporating and superheating in the two flow directions when multipass evaporators are considered.

The program computes the conditions of the two fluids at every node, describing the thermal and flow evolution inside the whole evaporator (Figure 104 on page 317).

The model was successfully validated on the basis of experimental tests carried out in our laboratory on medium and high capacity evaporators (Figure 105 on page 318).

Finally it was employed to design high efficiency evaporators for some manufacturers of heat pump equipments.

23.7 Steam Reforming Reactor for Fuel Cells

Project leader: F. Amodio

A computer code has been developed in order to evaluate the performances of a chemical reactor for the steam reforming of methanol; the hydrogen rich gas produced by the reactor will feed phosphoric acid fuel cells.

The one-dimensional equations used are based on following models:

- kinetics: model of Langmuir
- convective heat transfer in packed bed; model of Leva
- pressure gradient in packed bed: model of Leva

The model required the solution of a set of four ordinary differential equations; a second order accurate finite differences technique was employed for the discretization and the resulting tridiagonal set of algebraic equations was solved by means the simple Thomas procedure.

The experimental validation of the model showed deviations in the ranges of $\pm$ 5% between the measured and the calculated conversion factor.

23.8 Condensing Heat Exchangers.

Project Leader: M. Mattiello

It has been developed a computer code for the calculation and design of condensing heat exchangers to install at the exit to the stack of heating generators in order to recover from the flue gases the sensible heat and, below certain temperature, also the condensation heat from the water vapor in the gases.

This heat exchanger condenses the water vapor from a mixture with non condensing gases as CO_2, H_2 . O_2 and minor quantity of CO and $NO - NOx$.

The heat exchanger, depending on the temperature of the cooling fluid and on the vapor partial pressure, operates at three different conditions that are in sequence along the flue gas flow:

a) dry conditions without condensation
b) condensation regime with undersaturated flue gases in which the wall temperature of the flue gas side is below the dew point, while in the gaseous flow the vapor partial pressure is below the saturated one
c) condensation regime with saturared flue gases.

The model is unidimensional and is based an local transfer conditions for different geometries of cross flow tubes.

The program was validated experimentally for different geometries with good results. The difference in the calculation of the heat flow is maximum 7% and the temperature differences are below 10°C; this maximum differences are for the extreme working conditions of the heating generators.

23.9 Design of Natural Draft Boiler Sections

Project leader: M. Mattiello

Usually the cast iron boilers are composed assembling a number of sections which confine the combustion chamber and the convective heat channels.

A computer code for the evaluation of the convective and radiative heat transfer for different geometries of the sections has been developed.

The program evaluates at first the radiative heat transfer in the flame zone; the model of well stirred reactor has been used.

The calculation of the radiative heat transfer provides the temperature of the combustion gases at the convective channels, which are supposed to have the same mass flow rate.

Each channel is assumed to be formed by two parallel flat plates with symmetric series of spins and fins.

The convective heat transfer model is unidimensional.

The validation of the program has been done carried out comparing the following experimental data with the calculated ones:

- combustion gases temperatures at the the convective channels inlet
- combustion gases temperatures at the channels outlet
- heat flux from the gases.

The code has been validated for different geometries and at different heat rates for the same geometry; the agreement between the experimental and the calculated data is quite good, giving the maximum difference of $15°C$ for the outlet temperature of the gases, which is negligible in terms of heat rate.

The program was used to improve the efficiency of existing boiler sections, as well as to design new sections of different geometries for new boilers.

There is also a certain number of other programs used far different applications, (calculation of chimney, calculation of seasonal performance of heating or air conditioning appliances, etc.).

23.10 General Considerations

The trends in using mathematical modeling is toward a growth, for a research center as TECNARS, but there are some barriers against this that must be overcome.

The two main obstacles are the costs and the human resources: the industries, and particularly the small and medium size ones, are not willing to pay (or they cannot) the costs incurring in the development of computer models; their needs are usually for a fast and good answer to their problems, at a known cost.

They require results "good enough" from an engineering point of view, but not "scientifically" precise.

On the other hand, it is very difficult to estimate correctly the men-hours necessary to develop a physical and mathematical model, to generate the software, make it running properly, validate it, and only at this point use it for simulations.

In order to reduce the costs, a possible way is to develop more "generally applicable" computer models under contract from governmental research agencies, as the CNR - National Research Council or ENEA - National Agency for Alternative Energy; cooperation with University is also viable, but usually the bureaucratic aspects of contracts with state organizations are time consuming and exausting.

Other ways are the sharing of the costs through cooperation with other research organizations, and the purchasing of computer models available from specialized (profit) companies.

In the first case, TECNARS is participating in activities in the frame of the International Energy Agency and of the International Flame Research Foundation; we do believe that international cooperation is an effective way, if not the only one, to solve cost problems and to develop better performing models.

Acquiring existing computer models is an apparent fast way; it requires the understanding of the model, the validation of it and the correct application to the specific problem. The protection of the source program for commercial use is in contrast with this basic needs for the research engineers.

Human resources are not widely available; basic university courses not always include a sufficient level of experience in modeling techniques or in other cases they create high level specialists.

The consequence is the difficulty in the right approach to the technical problem within the research team and in the "communications" among the research engineers, the numerical method experts and the industry customer. It takes time to develop an "integrated" mentality.

In spite of this difficulties, mathematical methods will be more and more used as they are fundamental for the future. As an example, this is evident for the development of the advanced ceramic materials applications: any application require a wide thermal stress analysis using computers.

Organizations as our host, SASIAM, are in our opinion very well suited to overcame the cited difficulties; industrial mathematics, formation, software library, international links are both the objectives of SASIAM and the keys for the future.

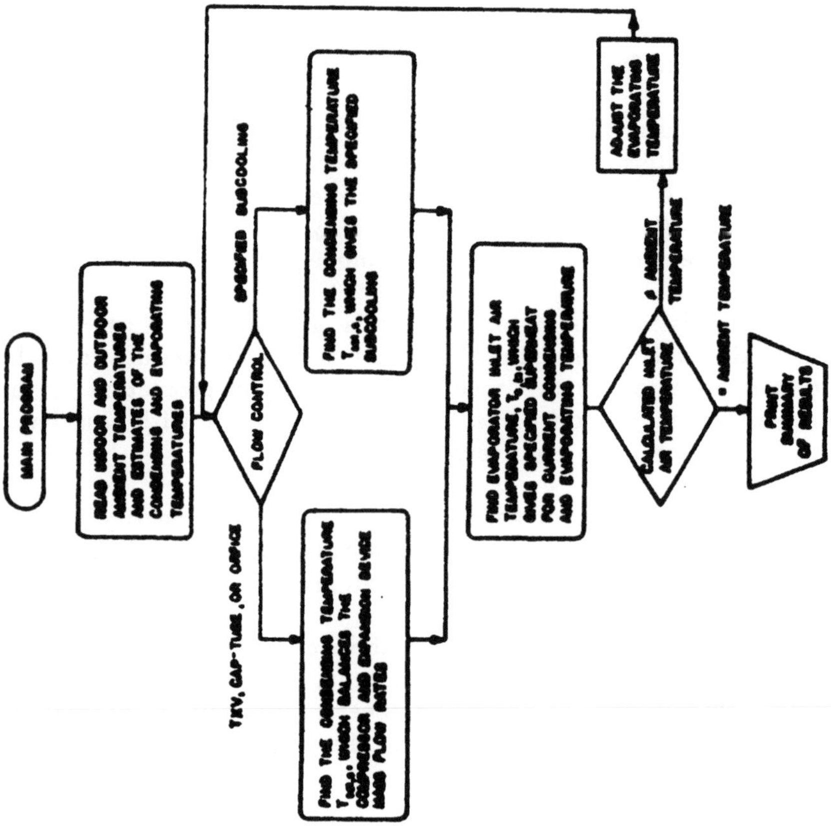

Figure 99. Block Diagram of Iterative Loops in the Main Program

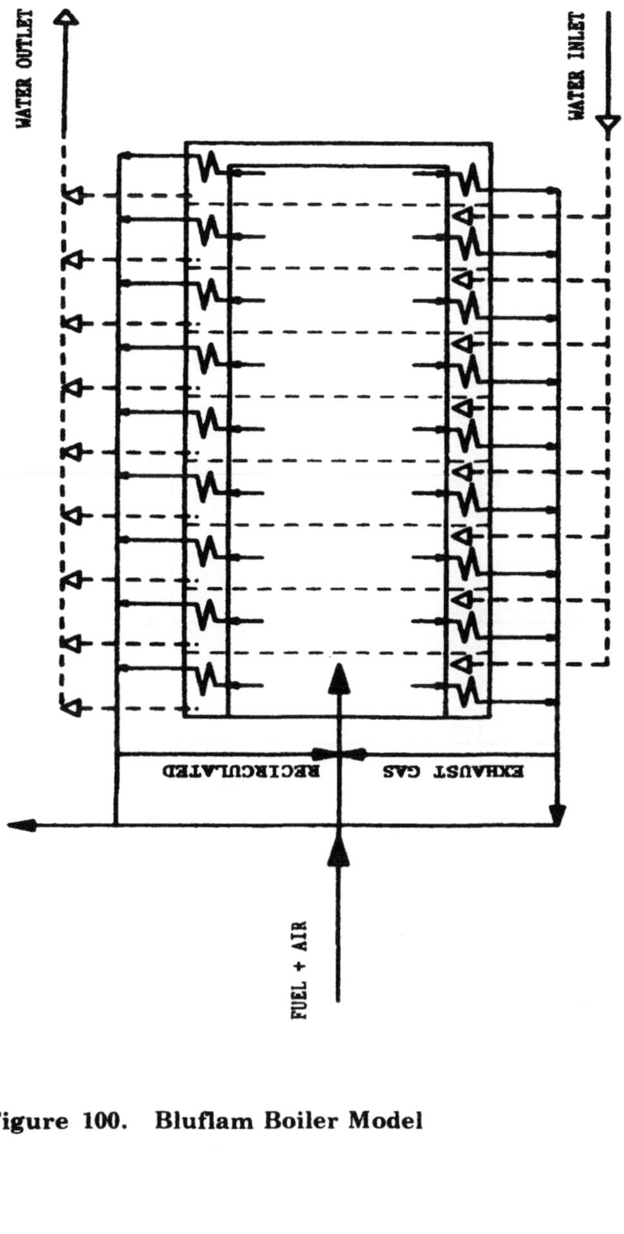

Figure 100. Bluflam Boiler Model

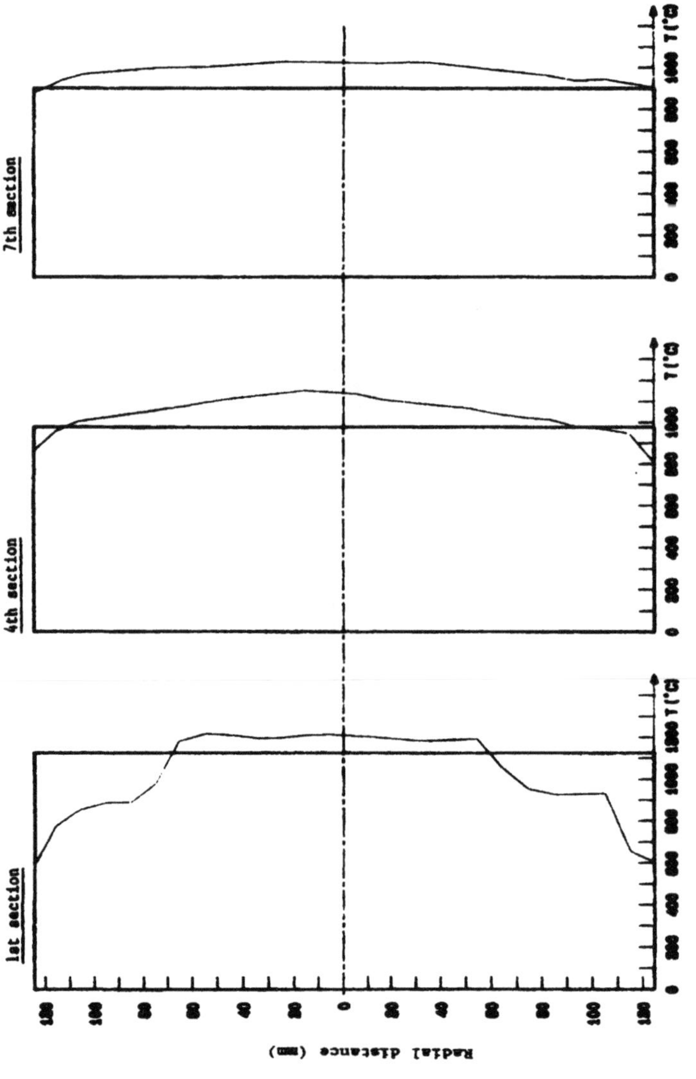

Figure 101. Experimental temperature profiles in three different sections of the chamber of a 9 section oil fueled bluflam boiler compared with the constant values predicted by the computer code

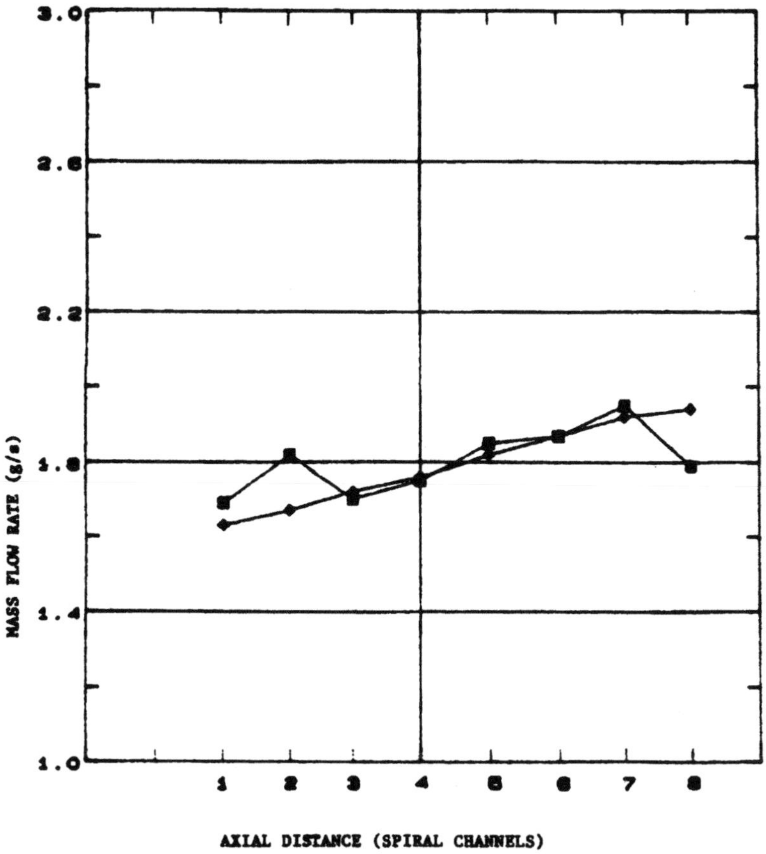

Figure 102. **Numerical (◇) and experimental (■) values of the combustion products mass flow rate in the spiral channel of a 9 sections oil fueled bluflam boiler**

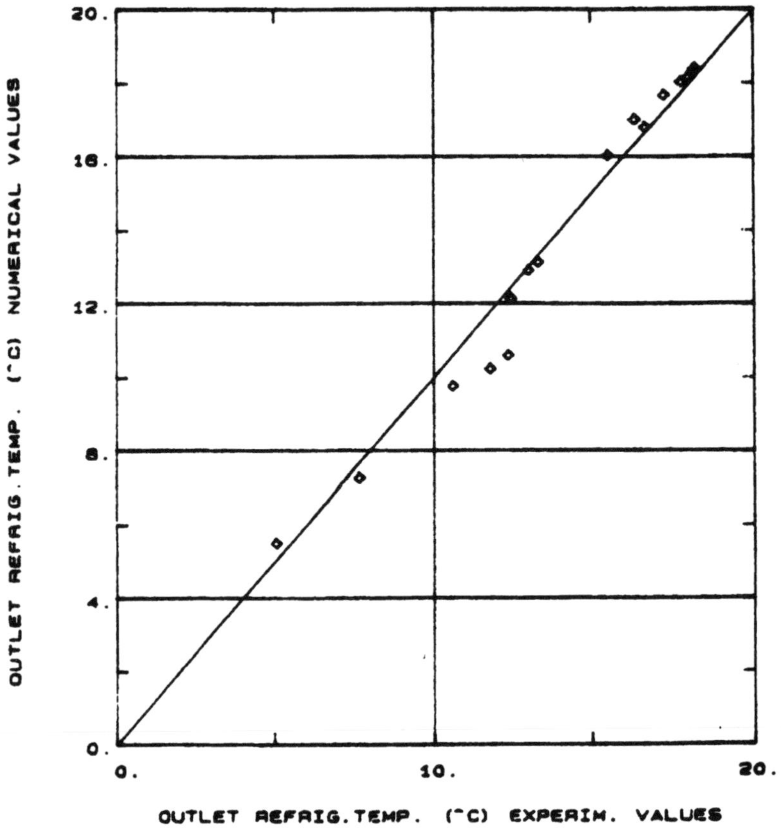

Figure 103. **Comparision between experimental values and numerical values of refrigerant temperatures at the evaporator exit. Counter-flow evaporator with refrigerant inside tubes**

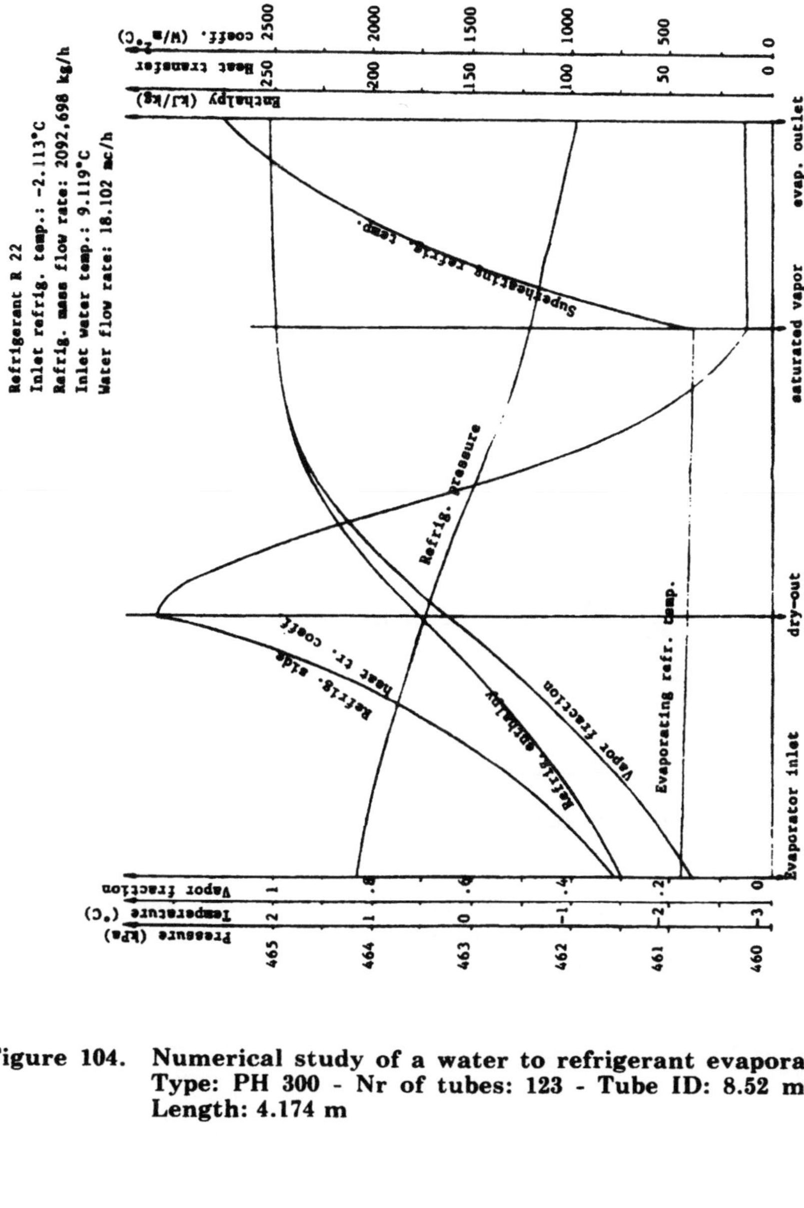

Figure 104. **Numerical study of a water to refrigerant evaporator. Type: PH 300 - Nr of tubes: 123 - Tube ID: 8.52 mm - Length: 4.174 m**

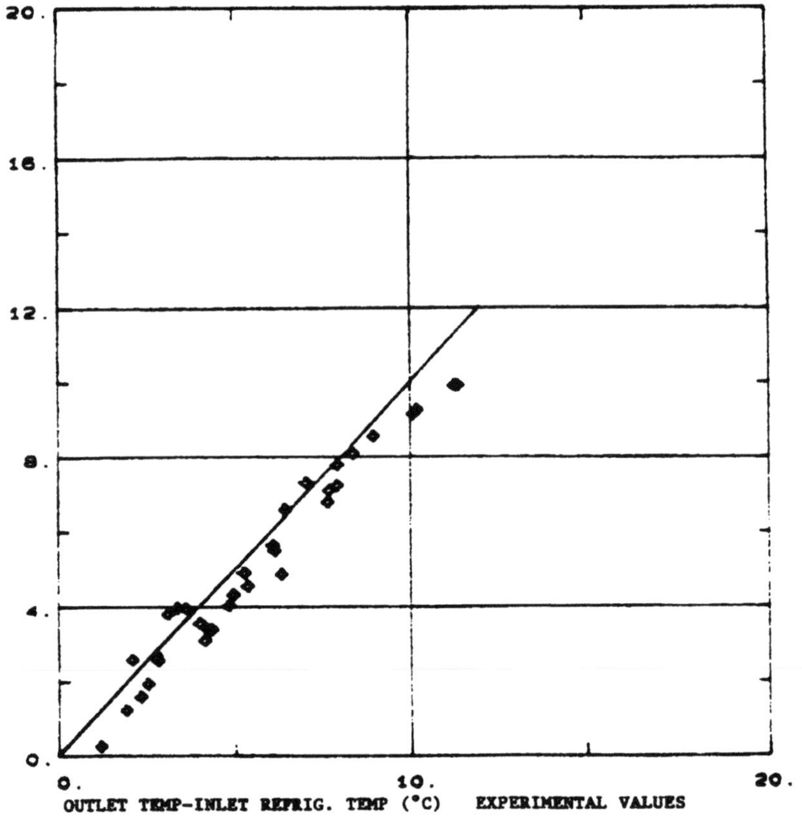

Figure 105. Comparision between experimental and numerical values for a shell-and-tube evaporator with refrigerant inside tubes

24. A Simulation Experience of a SMD Automatic Assembly Line

M. Orsucci

Telettra SpA - Vimercate

24.1 Abstract

Automatic assembly machines are becoming more and more usual in manufacture environments. The high flexibility of programmable tools make new problems to arise concerning the optimization of the machines potentialities.
Due to the complexity of the production problems, classical analytical methods can be hardly employed. So in many cases the solution is to simulate the time behavior of the system.

Simulation may be profitably employed as an economic and not destroying tool both in the design phase of the production line and in the operative phase of scheduling.

We shall present our experience in the study and implementation of a behavioral model of a SMD assembly line. The model is implemented in SIMSCRIPT.II language and is suitable nowadays for the tuning of the system and in the next future as part of the operative system of the line.

24.2 Introduction

This research experience takes its origin from a FAMOS project approved last year from the European Community. In this project Telettra is involved as one of the partners. The project, named I.D.E.A.M. from Integrated Design with Engineering and Automated Manufacturing, has the target of designing and developping a group of flexible automated workshops for the production of electronic cards, ranging from the preparation of naked printed boards to the quality control and testing of the cards.

The project includes the setting up of the assembly line, the guidance and management system and the automation of transfers between production stages.

In hardware and process data area are involved component performing, automated handling systems, test program generation from CAE/CAD systems, test area capture and analysis, automated handling at in-circuit test, odd-form components handling using robots.

In software area the main developments regards inventory management, master production schedule, optimization, simulation, resource management, cell management, area management.
Some degree of efforts includes the development of the local area network.

It is clear that the required complexity and the flexibility makes the system hardly to be handle with traditional methods.
Flexible manufacturing systems may be more productive and with less interruption than the rigid ones but they need an efficient automatic operating system.

Like big computers there is the case of many processes sharing common resources; the control software of the production plant must run like an operating system, optimizing the resources assignment in real time.
So special simulation languages, historically born for computers analysis, may be profitably employed.

As in most design-developments two phases are required: the synthesis and the analysis. Time-domain discrete-event simulation of complex system of coarse ownes to the second phase.
Others methods like queue networks analysis may be suitable for developping a first approximation solution to the problem, but only a simulation tool is able to represent the real situation as close as we need with the required flexibility. As it has no implicit limitations may be used both in the design optimization phase, to verify bugs and inefficiencies and in the operating system to help in the scheduling of the jobs, minimizing set-up changes while considering the present status, minimizing lead-time and WIP

24.3 The real situation

The first step is to describe the situation to be modelled. Figure 106 on page 328 represents in part the present situation of the Telettra SMD production line and in part as it is like to evolve in the next future thanks to the development of FAMOS project.

The main production phases may be recognized as solderprint, HS-180 and HS-72 pick and place machines, infrared oven, vapor phase solder, wash machine, wave solder, visual control, testing and repairing islands, robot for odd-form component placing.

The cable of the ETHERNET network can be recognized in the picture, and also terminal for controlling the incoming jobs, the line status and the generation and transfer of the part programs.

Moreover bar code readers will monitor the work in process on the line. In Figure 107 on page 329 a flow diagram is reported showing the working phases.

24.4 The methodology

The second step is to make a model of the real situation, that means to represent it with abstract entities and abstract formalized operations between them. There is a large amount of arbitrarieties in this procedure, because the complexity of a model depends closely on what we need to know about the reality itself. So the first thing to do is to have a sufficiently clear idea of the use we want to do of the model.

We can summarize the methodology adopted in developping the model as follows:

- Analysis of the real situation
- Definition of the main targets
- Construction of a model by introducing the needed hypothesis
- Codification of the model in a suitable computer language
- Verification that the model does work as we want
- Analysis of the input
- Confirming the validity of the model in representing the aspects of the reality that we want to control

- Design of the experiments and analysis of the results

The optimization of the system in order to achieve the best results, not automatically implemented, may be in some way performed with a suitable choice of the simulation parameters.

24.5 The targets

The main target is to have a control of parameters that are typical of many queuing systems such as

- Leadtime of the processes
- Work in process
- Rate of utilization of the machines
- Value of the queues in the buffers
- Production capability

In the hypothesis of a given rate of input jobs and scheduled work phases the benefit function of the system can be evaluated with different value of the system variables. Viceversa in the presence of a prefixed situation with a given system status the target may be to evaluate different scheduling strategies.

24.6 The model

As the parameters that we must evaluate are deriving from accumulation of discrete time information, it is clear that also the model and the simulation mode are likely to be discrete.
The entities forming the model may be divided into processes and resources.

The processes are temporarely entities that have a start and an end-time and employ or destroy resources and time in competition with other processes.

The resources are typically permanent entities that may be characterized by some attributes as a service-time, shareability etc.
The resources may be requested, relinquished or destroyed by processes and in function of their status they give origin to the queues.

The processes in the model are:

- The main process that acquires or generates the system inputs and gives origin to the other processes.
- The process "batch" represents a kind of job and is characterized by a quantity of job units and a work cycle. It generates a sequence of processes "rack".
- The process "rack" represents the generation of a packed fixed number of printed boards to be manufactured. It constitutes the job unit in our system and produces a computed working-time on every machine of which it makes a request.
- The process "set-up" is generated by the process "batch" when the job requests a modification of the machine preparation. It produces a set-up waiting time
- The process "interruption" is a random process that requires service from machines with the higher priority. Each "interruption" makes also request of one or more operators.

The resources in the model are:

- The "machines", characterized by attributes that modelize the typical parameter of the physical machines such as the working-time distribution function, the interruption-time distribution, the repair-time distribution, the list of operators to which they pertain, the priority, necessity of set-up and so on.
- The "Operators", characterized by a name, a logical status of business or availability, the name and the priority of the employing process.

The processes activity follows logic rules that try to reproduce the real behavior of the system.

Notice that many status or functions of the system are not modelled, e.g. the exact status of the machines set-up is not taken into account, but that agrees with the level of definition we want of the reality.

24.7 Analysis of the inputs

The analysis of the input is not less important than the analysis of the system and a misunderstanding of that is often cause of the failure of the entire simulation.

Typically real systems are subjected to input random variables that need to be described in statistical terms. Analysis of the inputs means to collect data, elaborate them statistically, make an analysis of the correlation between the random variables and compute the parameters of the distribution functions.

The analysis we describe here takes its origin from a three months observation of the input data. The more important data regard

- the time of set-up of the machines in function of the kind of job
- the working-time in function of the kind of job
- the time between two interruptions
- the availability of the operators.

Some random variables were found to have a dependence between them more or less foreseeable. In the first case a correlation method was adopted; in the second a method of clusterization.
The steps in the correlation analysis are :

- Collection of couples of data X Y

- Analysis of correlation by means of the coefficient of linear correlation

- Computation of the regression formulas using standard functions

- Choice of the best fit among the interpolating functions.

For example such analysis was performed in order to correlate the setup time to the number of component kinds on the board.
The correlation analysis showed some dependence between the two random variables. The regression analysis was performed using the following functions:

Linear : $Y = a + b \times X$

Exponential : $Y = a \times e^{(b \times X)}$

Power : $Y = a \times X^b$

Hyperbolic : $Y = a + b/X$

Reciproc : $Y = 1/(a + b \times X)$

Rational : $Y = X/(a + b \times X)$

These are classical formulas that have the advantage of being linear or directly or after an easy transformation. So the fit may be performed without iteration through the mean square method.
For each fit were computed the mean and the standard deviation of the error. The choice was made evaluating the best correlation factor between the observed and the interpolated data.

The clusterization method is used when the dependence of one random variable from another is not fairly described by a simple analytical law. Consider a cartesian diagram with the two axes representing the two random variables and draw a paint for every observed correspondence; if some points gather to form clouds it is possible to take one point to represent a group, obtaining in that manner a quantized distribution of the two random variables.
The centers of the clouds are found minimizing the sum of square errors.

Were the random variables may be considered independent, the problem is to find the best probability distribution or, in other words, to find the best estimator of the function depending on the sample of data. In our case the "maximum-likelihood estimator" were used.

In the hypothesis of "independent identically distributed" samples, the 'maximum-likelihood estimator" can be defined as the function

$$L(a_1, a_2, \ldots, a_K) = p\{x_1\} \times p\{x_2\} \times \ldots p\{x_N\}$$

with the set of parameters $a_1, a_2, \ldots, a_K$ that maximize it for the samples $x_1, x_2, \ldots, x_N$

As the extimation is made the problem is to verify the degree of confidence of it using a test of fit goodness.
The oldest goodness-of-fit hypothesis test is the chi-square test. To compute this test we must first divide the entire range of the fitted distribution into K adjacent intervals. Next we compute for each interval the observed number of events (N_j) and the expected ones (E_j). Finally, for discrete data, the test is defined

$$\chi^2 = \sum \left((N_j - E_j)^2 \frac{1}{E_j} \right)$$

This parameter is smaller as better is the fit. The chi-square test needs care in the use of it because it depends on the choice of the intervals, so we made use of the Kolmogorov-Smirnov test that compares an empirical distribution $G(x)$ function with the hypothesized distribution $F(x)$. The fitness of the test is measured simply by the larger vertical distance between $F(x)$ and $G(x)$ and is formally defined as

$$D = \sup \left(| F(x) - G(x) | \right)$$

and in the discrete case, with ordered samples X_j, we have

$$D = \sup \left(F(X_j) - \frac{j}{N} \right)$$

where N is the total number of samples. If the goodness of fit are not satisfactory, the last chance is to built an empirical histogram computing the relative frequency of the samples and to make use of it directly into the simulation.

24.8 Experiments and analysis of the outputs

After having validate the model together with the input statistics making use of the known history of the events, we could proceed to design experiments in order to control the system parameters and to verify the behavior of the system while varying something in it.

In order to compare the different hypothesized situations in the manufacture line, the following parameters were taken into account:

- Lead time, that is the mean time for a batch to go through the line.
- Completion time, that is the total time needed to complete all the batches.
- Absolute machine utilization time, that is the ratio between the working time of the machine and the completion time.
- Relative machine utilization time, that is the ratio between the working time of the machine and the total time that the machine is employed.
- Mean queue in number of printed boards.
- Mean queue in minutes.
- Rate of setup time, that is the ratio between the setup-time and the absolute utilization time of the machine.
- Rate of machine interruptions, that is the ratio between the interruption time and the absolute utilization time of the machine.
- Actual working time, of the batches respect to the completion time.

In Figure 108 on page 330 a table is reported showing an example of the results for two system configuration. This table is an objective tool of decision for evaluating the convenience of introducing other machines or operators.

In this case the main result is that including one more SIEMENS-180 and one more operator the completion time reduces almost to one half.

Moreover the solderprinter is not a bottleneck but it can cause a queue because it is the first phase of the cycle. This consideration may suggest some modification in the work organization.

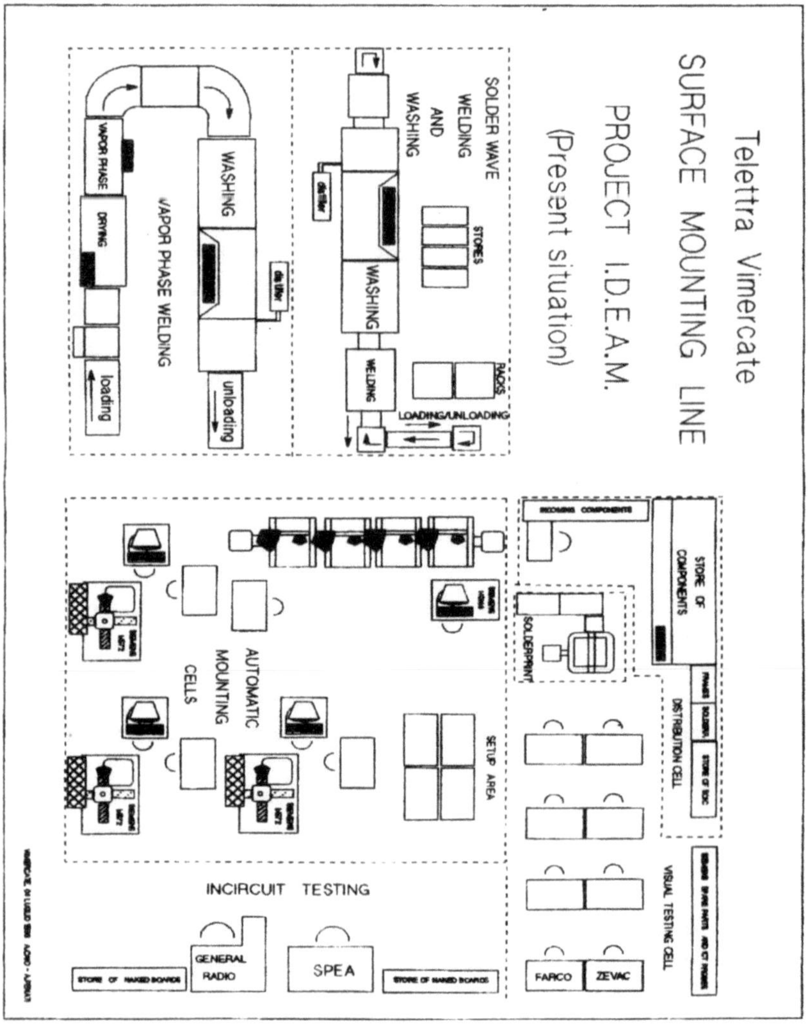
Telettra Vimercate
SURFACE MOUNTING LINE
PROJECT I.D.E.A.M.
(Present situation)
SOLDER WAVE WELDING AND WASHING
STORES
RACKS
WASHING
WELDING
LOADING/UNLOADING
VAPOR PHASE WELDING
VAPOR PHASE
DRYING
loading
unloading
AUTOMATIC MOUNTING CELLS
SETUP AREA
SOLDERPRINT
STORE OF COMPONENTS
DISTRIBUTION CELL
VISUAL TESTING CELL
INCIRCUIT TESTING
GENERAL RADIO
SPEA
FARCO
ZEVAC

Figure 106.

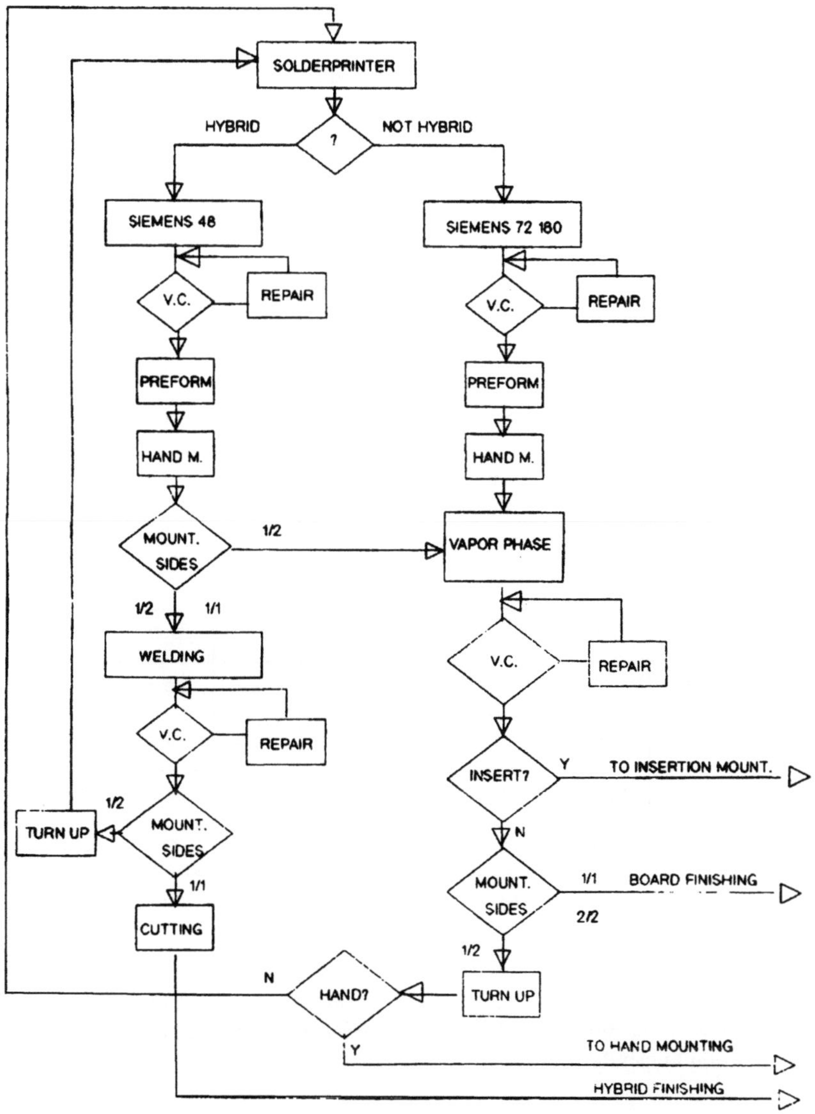
SOLDERPRINTER
HYBRID
?
NOT HYBRID
SIEMENS 48
SIEMENS 72 180
V.C.
REPAIR
V.C.
REPAIR
PREFORM
PREFORM
HAND M.
HAND M.
MOUNT. SIDES
1/2
VAPOR PHASE
1/2
1/1
WELDING
V.C.
REPAIR
V.C.
REPAIR
INSERT?
Y
TO INSERTION MOUNT.
N
1/2
TURN UP
MOUNT. SIDES
1/1
CUTTING
MOUNT. SIDES
1/1
BOARD FINISHING
2/2
1/2
N
HAND?
TURN UP
Y
TO HAND MOUNTING
HYBRID FINISHING

Figure 107.

RESULTS OF TWO SIMULATION RUNS

CASE 1 – COMPLETION TIME : 110106 UNITS

	AU	RU	QN	QT	SR	IR	WR
SOLDERPRINT	0.11	1	15.46	672	72.2	0	100
SIEMENS 72	0.98	1	96.40	13601	55.1	19.3	80.7
SIEMENS 180	0.99	1	0	0	20.0	29.2	70.8
WASHING	0.18	0.18	0.01	1	0	0	100
OPERATOR	0.91	0.92	0.71	194	0	0	100

CASE 2 – COMPLETION TIME : 61772 UNITS

	AU	RU	QN	QT	SR	IR	WR
SOLDERPRINT	0.20	1	27.17	1181	72.2	0	100
SIEMENS 72	0.97	1	86.85	79151	16.1	19.3	80.7
SIEMENS 180-1	0.99	1	0	0	11.8	29.1	70.9
SIEMENS 180-2	0.98	1	0	0	11.9	29.2	70.8
WASHING	0.31	0.32	0.04	3	0	0	100
OPERATOR-1	0.77	0.79	0.20	58	0	0	100
OPERATOR-2	0.86	0.86	0.20	58	0	0	100

AU : Absolute utilization
RU : Relative utilization
QN : Queue in number of boards
QT : Queue in minutes
SR : Setup time rate
IR : Interruption time rate
WR : Effective working time rate

Figure 108.

Acknowledgements

It is our privilege to thank the many people who provided moral support, financial assistance, and technical help in connection with the International Workshop on the Mathematical Methods for Industrial Problems, held in Tecnopolis, Bari, Italy in September 1988.

The conference was made possible thanks to the support of the following institutions:

* Consiglio Nazionale delle Ricerche - Comitato Nazionale per le Scienze Matematiche
* FORMEZ - Centro di Formazione e Studi per il Mezzogiorno
* Regione Puglia - Assessorato al Lavoro, alla Cooperazione e alla Formazione Professionale
* Tecnopolis CSATA Novus Ortus

The Workshop was held under the auspices of ECMI (European Consortium for Mathematics in Industry).

Special thanks are due to Dr. Ing. Umberto Bozzo, Director General of Tecnopolis CSATA Novus Ortus, who has always supported, through SASIAM, the role of Mathematics in technological development. We also wish to thank the whole staff of Tecnopolis CSATA Novus Ortus who directly or indirectly contributed to the success of this initiative.

We finally wish to express all our appreciation to the staff of the Electronic Publishing Laboratory of Tecnopolis CSATA Novus Ortus and in particular Dr. G. Storelli for their valuable contribution in the preparation of this volume.

V. CAPASSO
R. CASELLI

Bei Fragen zur Produktsicherheit wenden Sie sich bitte an:
If you have any questions regarding product safety,
please contact:

Walter de Gruyter GmbH
Genthiner Straße 13
10785 Berlin
productsafety@degruyterbrill.com